AF356455

Rheology and Quality Research of Cereal-Based Food

Rheology and Quality Research of Cereal-Based Food

Editors

Anabela Raymundo
María Dolores Torres
Isabel Sousa

MDPI • Basel • Beijing • Wuhan • Barcelona • Belgrade • Manchester • Tokyo • Cluj • Tianjin

Editors
Anabela Raymundo
Universidade de Lisboa
Portugal

María Dolores Torres
Universidade de Vigo
Spain

Isabel Sousa
Universidade de Lisboa
Portugal

Editorial Office
MDPI
St. Alban-Anlage 66
4052 Basel, Switzerland

This is a reprint of articles from the Special Issue published online in the open access journal *Foods* (ISSN 2304-8158) (available at: https://www.mdpi.com/journal/foods/special_issues/Rheology_Quality_Cereal-Based_Food#).

For citation purposes, cite each article independently as indicated on the article page online and as indicated below:

LastName, A.A.; LastName, B.B.; LastName, C.C. Article Title. *Journal Name* **Year**, *Volume Number*, Page Range.

ISBN 978-3-0365-0504-6 (Hbk)
ISBN 978-3-0365-0505-3 (PDF)

Cover image courtesy of Pedro Raymundo Carlota.

Contents

About the Editors

Anabela Raymundo is a chemical engineer with an M.Sc. in Food Science and Technology and a Ph.D. in Food Engineering. She is an Assistant Professor with Habilitation whose activities include integrating of LEAF (Linking Landscape, Environment, Agriculture and Food) and responsibility in the areas of Rheology and Food Texture and Quality Control in Food Engineering in graduate and master programs, such as the Masters of Food Engineering and Gastronomic Sciences. Her main work is focused on the use of poorly exploited food sources (e.g., microalgae biomass and food industry byproducts) for the development of products with high added value.

Maria Dolores Torres' research interests are focused on the valorization of byproducts from the agri-food industry and the development of innovative functional gelled foodstuffs mainly directed toward target population groups with special nutritional requirements. Currently, she is working on the integral valorization of brown and red algae using environmental friendly technologies to obtain functional hydrogels and other components of industrial value.

Isabel Sousa, Ph.D. in Food Science from University of Nottingham, UK, is Associate Professor with Habilitation at Instituto Superior de Agronomia (ISA), Universidade de Lisboa. She is Head of the LEAF (Linking Landscape Environment Agriculture and Food) research center at the Portuguese Foundation for Science and Technology (FCT) and, in Portugal, an expert and pioneer of Food Texture and Rheology research and teaching. In addition, she is the coordinator of numerous national and European projects with SMEs.

Editorial

Special Issue: Rheology and Quality Research of Cereal-Based Food

Anabela Raymundo [1],*, María Dolores Torres [2] and Isabel Sousa [1]

[1] LEAF—Linking Landscape, Environment, Agriculture and Food, Research Center of Instituto Superior de Agronomia, Universidade de Lisboa, Tapada da Ajuda, 1349-017 Lisboa, Portugal; isabelsousa@isa.ulisboa.pt

[2] Department of Chemical Engineering, Science Faculty, Universidade de Vigo (Campus Ourense), As Lagoas, 32004 Ourense, Spain; matorres@uvigo.es

* Correspondence: anabraymundo@isa.ulisboa.pt; Tel.:+351-21-365-2114

Received: 12 October 2020; Accepted: 19 October 2020; Published: 22 October 2020

Abstract: New trends in the cereal industry deal with a permanent need to develop new food products that are adjusted to consumer demands and, in the near future, the scarcity of food resources. Sustainable food products as health and wellness promoters can be developed redesigning traditional staple foods, using environmentally friendly ingredients (such as microalgae biomass or pulses) or by-products (e.g., tomato seeds) in accordance with the bioeconomy principles. These are topics that act as driving forces for innovation and will be discussed in the present special issue. Rheology always was the reference discipline to determine dough and bread properties. A routine analysis of cereal grains includes empirical rheology techniques that imply the use of well-known equipment in cereal industries (e.g., alveograph, mixograph, extensograph). Their parameters determine the blending of the grains and are crucial on the technical sheets that determine the use of flours. In addition, the structure of gluten-free cereal-based foods has proven to be a determinant for the appeal and strongly impacts consumers' acceptance. Fundamental rheology has a relevant contribution to help overcome the technological challenges of working with gluten-free flours. These aspects will also be pointed out in order to provide a prospective view of the relevant developments to take place in the area of cereal technology.

Keywords: gluten-free products; dynamic oscillatory shear measurements; pasting profile; rheology; texture; X-ray diffraction; red kidney bean; microalga *Tetraselmis chuii*; acorn flour; tamarind gum; yogurt; tomato seed flour; antioxidants; phenolics; fibre-rich ingredient; ball milling

1. Introduction

Innovation plays a key role in the current development of food companies. Food trends are launched every year, allowing stakeholders to be aligned around common axes on consumer's preferences.

The most recent trends in the food industry [1] are focused on foods of vegetable origin, and commitment to sustainability stands out. The increase in the trend of personalized diet ("good for me") is also noteworthy, which includes gluten-free products or foods with a direct impact on health and well-being (rich in bioactive compounds).

It is well known that products with the same chemical composition can present very different structures that are built up by processing techniques, resulting in differently perceived texture and sensory properties. Rheology has been the reference discipline for the food cereal industries since the very start of quality control. Food macromolecules (proteins and polysaccharides) are the major players for the creation of relevant food structures, such as dough and crisp snacks. The development of gluten-free products using alternative proteins and polysaccharides, nutritious mixtures of cereals and pulses to replace meat protein, as well as the use of food industry by-products, such as tomato seeds as a source of these structuring biopolymers (pectins), are some other challenges in creating innovative food products. Sustainability in the production of the food ingredients and the economic viability of their production and subsequent transformation into fair-traded, well-accepted food products, are essential for the progress of the cereal industry, with a relevant impact on human wellness and progress. The use of poorly exploited food sources, such as microroalgae, in cereal products is also an opportunity to explore. This approach includes the design of added value food products and relevant nutritional benefits. The use of ancient flours that have fallen out of use, such as acorn flour or ancient wheat varieties with low-gluten proteins, could be a sustainable strategy to enhance cereal products with an important source of bioactivities and fiber.

Finally, consumer attitude toward new food products is a relevant issue for the success of innovations and should be considered for food products that are close to the market. In this sense, the sensory evaluation of innovative products, in the preliminary stage of the development process, is an important tool in order to predict their final acceptance of the market.

2. Contributions

Gluten-free foods stand out in terms of food trends in the field of cereal technology. These types of products are on the agenda of the most important food companies and many research groups. This results from the steady increase of the gluten-free products market due to the growing number of individuals diagnosed with some type of gluten sensitivity. About 38% of consumers are avoiding or limiting gluten-free foods [2].

Generally, gluten-free foods are nutritionally unbalanced in terms of lipids, fibers, minerals, and vitamins. This can be critical for the celiac patients, who have co-pathologies and therefore need a nutritionally balanced diet [3].

As bakery products are traditionally produced with gluten flours, they have undergone extensive redesign in order to optimize their production from gluten-free flours. In this context, bread takes on a considerable prominence. Several works on this theme have been published in recent years [4–6], which are related to the incorporation of underexplored ingredients and by-products of the food industry, in order to improve the rheological and texture properties, nutritional performance, and sensory appreciation. In the present special issue, relevant studies related to gluten-free products are present.

Nunes et al. (2020) [7] evaluated the impact of *Tetraselmis chuii* microalgae incorporation on the structure, color, and bioactivity of a gluten-free (GF) bread formulation based on rice, buckwheat flour, and potato starch to increase nutrition, using hydroxipropylmethylcellulose (HPMC) as a structuring agent. The dough pasting profile assessed by Microdough-Lab and shear oscillatory measurements was conducted to evaluate the dough structure. Physical properties of the loaves, total phenolic content (*Folin–Ciocalteu*), and antioxidant capacity (DPPH and FRAP methods) of the bread extracts were assessed. Promising results related with the use of *T. chuii* as a sustainable ingredient in GF bread formulation were found, with a positive impact on the structure and antioxidant capacity and an innovative green appearance. This microalgae presented different behavior, according to the incorporation level: below 2%, *T. chuii* proteins destabilize the structure developed by starch and HPMC, and smaller bread volume was obtained, which was associated with a more compact crumb and harder properties. However, for higher levels of incorporation (4%), the microalgal proteins with starch and HPMC build up another type of structure, which is characterized by higher values of the viscoelastic functions (G′ and G″) producing higher bread volume and a softening effect. There was evidence that the structure of 4% *T. chuii* bread is competitive with the control bread (with no biomass addition), in terms of structure and with a boost in nutritional performance, despite having revealed a weak acceptance by a non-targeted sensory panel.

Martins et al. (2020) [8] studied the possibility of using acorn flour, an under-exploited GF raw material, in a similar formulation as the one developed by Nunes et al. (2020). This flour was tested in order to improve dough rheology, following also market trends of sustainability and fiber-rich ingredients. Acorn flour significantly affected the rheology properties of the doughs. An impact on the dough's mixing and pasting curves, an improvement of the texture parameters (firmness and cohesiveness) and the viscoelasticity of the fermented dough were highlighted. In this way, the role of dough characterization by rheology tools was evidenced, as being determinant for the optimization of new food products. According to small amplitude oscillatory shear measurements, all the GF doughs studied exhibit a weak gel-elastic-like behavior with G' values higher than G'' and frequency dependent. Acorn flour incorporation caused the acidification and increased the darkness of the dough, which can have a positive impact in terms of sensory appreciation of the bread. Therefore, it was stated that acorn flour can be a very promising ingredient to improve both the rheological GF dough properties and nutritional GF bread quality, in particular dietary fiber content, which is a really important nutrient in special requirement diets.

In line with the work already presented on the nutritional enrichment of GF bread, Graça et al. (2020) [9] studied the possibility of enriching a similar type of bread with yogurt. Following this strategy, the low functional and nutritional properties of GF bread can be minimized, using dairy protein. Fresh yogurt represents an interesting ingredient since in addition to being an important protein source, it also provides polysaccharides and minerals that have the potential to mimic the gluten network, while improving the nutritional value of gluten-free products. Gluten-free bread formulations, with different levels of yogurt addition (5% up to 20%, *weight/weight*), were evaluated, using dough rheology measurements and baking quality parameters. It was shown that the functionality of gluten-free breads, in terms of bread-making performances, quality parameters, and nutritional profile can be successfully improved by the addition of fresh yogurt, resulting in a significant improvement in the overall quality of the GF yogurt-breads. Linear correlations between bread firmness, specific volume with flow behavior, and viscoelastic functions were found, supporting the results obtained. These correlation can assume a considerable importance in terms of the bakery industry and for future studies in this area. Yogurt was shown to be a potential ingredient to improve the quality of gluten-free breads, resulting in softer breads with higher volume and lower staling rate, compared to control bread. In relation to the nutritional composition, yogurt addition was revealed to be an attractive ingredient to enhance the nutritional value of GF breads: an increase in the protein and mineral contents and a reduction in carbohydrates was found, with a good chance to improve the daily diet of celiac people.

Hong and Kweon (2020) [10] presented another study on gluten-free rice bread, incorporating tamarind gum. In this work, the importance of optimizing the formulation and processing conditions is highlighted when combining new ingredients for product design. An experimental factorial design was used and revealed to be useful for the optimization process, minimizing the number of experiments and emphasizing the weight of each independent variable in the explanation of the process. Gum concentration (GC), water amount (WA), mixing time (MT), and fermentation time (FT) were selected as factors, and two levels were used for each factor. WA and FT were identified as the most significant factors to determine the quality of GF rice bread with tamarind gum. Therefore, proper control over the water content and fermentation time can maximize the bread volume and minimize the firmness of the bread. The addition of an anti-staling enzyme was also studied and proved to be effective in retarding the retrogradation enthalpy and decrease of bread firmness. Using an optimized formula and processing factors for gluten-free rice bread with the combined addition of tamarind gum and an anti-staling enzyme (maltogenic amylase) can be applied successfully in commercially manufactured gluten-free rice bread.

In addition to gluten-free bread, several formulations of bakery products, based on GF flours, have appeared on the market. Chompoorat et al. (2020) [11] studied the impact of rice flour addition on red kidney bean (RKB) cupcakes. The incorporation of rice flour promoted an increase in the degree of structuring of the cupcake dough and an improvement in the texture properties of the RKB cupcakes. It is important to note that also for this type of matrix, the use of fundamental rheological techniques such as dough linear viscoelastic behavior and empirical tests such as texture characterization were crucial to optimizing the final product. Rice flour addition in gluten-free RKB flour increased the batter's solid-like and viscous behavior, batter consistency, inflection of gelatinization and temperature, and produced a softer cupcake texture. The activation energy of gelatinization also increased with rice incorporation, as well as the cupcakes' macrostructural characteristics. The potential of RKB as a functional ingredient and its improvement in cupcake application with the addition of rice flour was highlighted.

Arribas et al. (2019) [12] studied another relevant GF product: an extruded rice snack. This type of product, in addition to being part of the current trend in the consumption of GF cereal foods, assumes an important position in the snacking food trend [13]. Snacking has intensively grown in the last few years and is associated with the growth of new forms of consumption associated with more dynamic lifestyles. The authors evaluated the impact of adding two legumes, bean and carob fruit flours, on the physical properties and bioactivity of GF puffed snacks. The fortification with carob fruit flour improved their textural attributes and did not significantly affect their overall quality. The extrusion had positive implications in terms of the nutrition and availability of bioactive compounds, as well as good acceptance in sensory terms. All the experimental extrudates had higher amounts of bioactive compounds than the commercial extruded rice. This process affected phytochemicals to a different extent. While total α-galactosides and phenols increased, the phytic acid was reduced, and the lectins and protease inhibitors were eliminated. The content of bioactive compounds present in these extrudates might be enough to promote health-associated functions. Moreover, the absence of lectins and protease inhibitors enhanced the nutritional quality of the extrudates. The developed snacks would be of interest to both health-conscious consumers and the food industry.

The development of gluten-free bakery products is a major challenge, which can be overcome through several strategies that were already presented in the works mentioned above. The use of hydrocolloids that are capable of creating a structure that mimics the gluten matrix is often followed. Nuvoli et al. (2020) [14] studied the effect of ball milling treatment on different types of hydrocolloids in a corn starch–rice flour system. Guar gum (GG), tara gum (TG), and methylcellulose (MC) were the hydrocolloids studied, and they were previously analyzed to assess their potential interactions with starch components, when they are used alone or in blends in a corn starch–rice flour system. Based on X-ray diffraction (XRD) experiments and gelling rheology characterization, it could be stated that the ball milling treatment affected the structure of the tested hydrocolloids and, in turn, the viscosity of their aqueous solutions in different ways. In fact, ball milling caused a reduction in the crystallin domain and, in turn, a diminished viscosity of the GG aqueous solutions. Despite an increase in its flow properties (viscosity), effects on TG were minimal, while the milled MC exhibited reduced crystallinity but similar viscosity. When both milled and un-milled hydrocolloids were individually added to the starch–flour system, the pasting properties of the resulting mixtures seemed to be affected by the type of hydrocolloid added, rather than by the structural changes induced by the treatment. All hydrocolloids increased the peak viscosity of the binary blends (especially pure GG). However, only milled and un-milled MC showed values of setback and final viscosity similar to those of the individual starch. Ball milling seemed to be more effective when two combined hydrocolloids (milled GG and MC) were simultaneously used. No significant differences were observed in the viscoelastic properties of the blends, except for un-milled GG/starch, milled TG/starch, and milled MC/milled TG/starch gels. The work presented was considered by the authors as an initial study, recognizing the need to deepen the theme. In this way, in future studies, the relevance of using the ball milling treatment, in the development of gluten-free bakery products, using hydrocolloids–starch systems will be clarified.

The enrichment of bakery goods, with and without gluten, by means of the incorporation of by-products from the food industry, has assumed special relevance, in recent years, taking into account the principles of the circular economy. Mironeasa and Codina (2019) [15] studied the utilization of a very relevant by-product of the food industry: tomato seed flour (TSF) produced. This work also highlighted the importance in the rheological and microstructure characterization of the dough on the bakery products development process. Rheology methods through the Mixolab device, dynamic rheology, and epifluorescence light microscopy (EFLM) were used to characterize the dough obtained from different formulations. From the Mixolab results, it was noticed that replacing wheat flour with

TSF increased the dough development time, stability, and viscosity during the initial heating–cooling cycle and decreased alpha amylase activity. The dynamic rheological data showed that the viscoelastic moduli (G′ and G″) increased with the level of TSF addition. Creep recovery tests of the samples indicated that the dough elastic recovery was at a high percentage after stress removal for all the samples in which TSF was incorporated in wheat flour. Using EFLM, all the samples seemed homogeneous showing a compact dough matrix structure. The parameters measured with Mixolab during mixing were in agreement with the dynamic rheological data and in accordance with the EFLM structure images. The authors show a correlation between the mixing properties and the viscoelastic behavior for the studied system (with gluten), as was also verified by Graça et al. (2020) [6], of GF bread.

The results presented in the different works of this special issue are useful for bakery producers to develop new products with the highest nutritional value, respecting the major key trend of sustainability and aligning with the major food trend to answer the consumer's actual needs.

Author Contributions: A.R.—writing and organizing of the document; I.S. and M.D.T.—writing and revised the document. All authors have read and agreed to the published version of the manuscript.

Funding: A.R. and I.S. are grateful to FCT for the financial support of LEAF—UIDP/04129/2020. M.D.T. thanks Spanish Ministry of Science, Innovation and Universities for her postdoctoral grant (RYC2018-024454-I).

Conflicts of Interest: The authors declare no conflict of interest.

References

1. EUFIC. Food Choice—Why Do We Eat What We Eat. Available online: https://www.eufic.org/en/food-safety/healthy-living/category/food-choice (accessed on 20 October 2020).
2. Hendry, N. *Innovation Opportunities across the Global Food Landscape*; Global Data Report; Portugal Foods: Lisbon, Portugal, 2019.
3. Missbach, B.; Schwingshackl, L.; Billmann, A.; Mystek, A.; Hickelsberger, M.; Bauer, G.; König, J. Gluten-free food database: The nutritional quality and cost of packaged gluten-free foods. *PeerJ* **2015**, *3*, e1337. [CrossRef] [PubMed]
4. Wang, K.; Lu, F.; Li, Z.; Zhad, L.; Han, C. Recent developments in gluten-free bread baking approaches: A review. *Food Sci. Technol. Campinas.* **2017**, *37*, 1–9. [CrossRef]
5. Beltraão-Martins, R.; Gouvinhas, I.; Nunes, M.C.; Peres, J.; Raymundo, A.; Barros, A. Acorn Flour as a Source of Bioactive Compounds in Gluten-Free Bread. *Molecules* **2020**, *25*, 3568. [CrossRef] [PubMed]
6. Graça, C.; Fradinho, P.; Sousa, I.; Raymundo, A. Impact of *Chlorella vulgaris* on the rheology of wheat flour dough and bread texture. *LWT—Food Sci. Technol.* **2018**, *89*, 466–474. [CrossRef]
7. Nunes, M.C.; Fernandes, I.; Vasco, I.; Sousa, I.; Raymundo, A. *Tetraselmis chuii* as a Sustainable and Healthy Ingredient to Produce Gluten-Free Bread: Impact on Structure, Colour and Bioactivity. *Foods.* **2020**, *9*, 579. [CrossRef] [PubMed]
8. Beltrão-Martins, R.; Nunes, M.C.; Ferreira, L.M.; Peres, J.R.N.A.; Barros, A.I.; Raymundo, A. Impact of Acorn Flour on Gluten-Free Dough Rheology Properties. *Foods* **2020**, *9*, 560. [CrossRef] [PubMed]
9. Graça, C.; Raymundo, A.; Sousa, I. Yogurt as an Alternative Ingredient to Improve the Functional and Nutritional Properties of Gluten-Free Breads. *Foods* **2020**, *9*, 111. [CrossRef] [PubMed]
10. Hong, Y.-E.; Kweon, M. Optimization of the Formula and Processing Factors for Gluten-Free Rice Bread with Tamarind Gum. *Foods* **2020**, *9*, 145. [CrossRef] [PubMed]
11. Chompoorat, P.; Kantanet, N.; Hernández Estrada, Z.J.; Rayas-Duarte, P. Physical and Dynamic Oscillatory Shear Properties of Gluten-Free Red Kidney Bean Batter and Cupcakes Affected by Rice Flour Addition. *Foods* **2020**, *9*, 616. [CrossRef] [PubMed]
12. Arribas, C.; Cabellos, B.; Cuadrado, C.; Guillamón, E.; Pedrosa, M. Bioactive Compounds, Antioxidant Activity, and Sensory Analysis of Rice-Based Extruded Snacks-Like Fortified with Bean and Carob Fruit Flours. *Foods* **2019**, *8*, 381. [CrossRef] [PubMed]
13. Mordor Intelligence. Free-From Food Market—Growth, Trends, and Forecast (2020–2025). Available online: https://www.mordorintelligence.com/industry-reports/free-from-food-market (accessed on 20 October 2020).

14. Nuvoli, L.; Conte, P.; Garroni, S.; Farina, V.; Piga, A.; Fadda, C. Study of the Effects Induced by Ball Milling Treatment on Different Types of Hydrocolloids in a Corn Starch–Rice Flour System. *Foods* **2020**, *9*, 517. [CrossRef] [PubMed]

15. Mironeasa, S.; Codină, G.G. Dough Rheological Behavior and Microstructure Characterization of Composite Dough with Wheat and Tomato Seed Flours. *Foods* **2019**, *8*, 626. [CrossRef] [PubMed]

Publisher's Note: MDPI stays neutral with regard to jurisdictional claims in published maps and institutional affiliations.

Article

Tetraselmis chuii as a Sustainable and Healthy Ingredient to Produce Gluten-Free Bread: Impact on Structure, Colour and Bioactivity

Maria Cristiana Nunes *, Isabel Fernandes, Inês Vasco, Isabel Sousa and Anabela Raymundo

LEAF—Linking Landscape, Environment, Agriculture and Food, Instituto Superior de Agronomia, Universidade de Lisboa; Tapada da Ajuda, 1349-017 Lisbon, Portugal; icxfernandes@gmail.com (I.F.); inesfsvasco@gmail.com (I.V.); isabelsousa@isa.ulisboa.pt (I.S.); anabraymundo@isa.ulisboa.pt (A.R.)
* Correspondence: crnunes@gmail.com; Tel.: +351-21-365-3100

Received: 31 March 2020; Accepted: 1 May 2020; Published: 4 May 2020

Abstract: The objective of this work is to increase the nutritional quality of gluten-free (GF) bread by addition of *Tetraselmis chuii* microalgal biomass, a sustainable source of protein and bioactive compounds. The impact of different levels of *T. chuii* (0%—Control, 1%, 2% and 4% *w/w*) on the GF doughs and breads' structure was studied. Microdough-Lab mixing tests and oscillatory rheology were conducted to evaluate the dough´s structure. Physical properties of the loaves, total phenolic content (*Folin-Ciocalteu*) and antioxidant capacity (DPPH and FRAP) of the bread extracts were assessed. For the low additions of *T. chuii* (1% and 2%), a destabilising effect is noticed, expressed by lower dough viscoelastic functions (G' and G") and poor baking results. At the higher level (4%) of microalgal addition, there was a structure recovery with bread volume increase and a decrease in crumb firmness. Moreover, 4% *T. chuii* bread presented higher total phenolic content and antioxidant capacity when compared to control. Bread with 4% *T. chuii* seems particularly interesting since a significant increase in the bioactivity and an innovative green appearance was achieved, with a low impact on technological performance, but with lower sensory scores.

Keywords: gluten-free bread; microalga *Tetraselmis chuii*; rheology; texture; colour; antioxidants; phenolics

---

1. Introduction

This study is part of Algae2Future project, that intends to explore the microalgae potential to be a low-carbon/nitrogen-footprints healthy food ingredient. These photosynthetic unicellular organisms have a huge importance in terms of the carbon dioxide mitigation [1] and nitrogen balance [2]. Moreover, microalgae are considered to be one of the most promising sources of functional food ingredients since their natural encapsulated bioactive compounds to promote important health benefits [3,4]. In the near-future context of food shortage and urge for sustainability, when the rate of population increase will be higher than increase of food production, caused by several environmental, social and economic factors, the use of alternative or under-exploited sources of protein is a very important issue. Microalgae are exceptional protein resources with potential to become a staple food for consumers all over the planet [1,5,6].

When considering microalgae as food for the future, it is also important to highlight that its incorporation into food is a challenge. There is a technological limit of microalgae incorporation, resulting from its impact on the food structure, that can be followed by a change on the rheology behaviour [7–9]. The introduction of microalgae biomass imparts changes in foods structure, but also in colour and flavour. Consumers are very sensitive to the changes in sensory characteristics, which induces limitations on the level of microalgae incorporation. *Tetraselmis chuii* is a green microalga approved by EFSA (Regulation UE 2017/2470) which has a high protein content, that is an important requirement to be used in bread with a specific nutritional profile. In the last few years, several works have been developed about the incorporation of microalgae into food. Many innovative food systems enriched with algae have been proposed, namely pasta [10–13], cookies [14–16] and wheat bread [17,18].

Bread is a staple food with specific characteristics in terms of the development of a cohesive and elastic dough structure. Since gluten confers unique rheological properties to yeast-leavened baked products, the absence of gluten is a major technological drawback. The combination of structural ingredients, including starches, gluten-free (GF) flours, hydrocolloids, proteins and additives, to develop an alternative to gluten dough´s structure has been widely tested [19–21]. Therefore, the addition of microalgae with high protein content can be important for the development of GF bread. In addition, nutritional benefits can be achieved by the addition of microalgae, and this is particularly important in GF bread, since celiac patients have nutritional deficiencies due to their absorption limitations.

Gluten-free is a current hot topic in the food industry. Consumption of GF products, and particularly bread, has increased considerably in recent times. This is due not only to the increase of celiac disease, but also to an increase in other gluten-related disorders [21]. Now, an increasing number of consumers, who have not been diagnosed with celiac disease, are cutting gluten from the diet and GF products are aligned with the top ten food trends.

Research on GF products with algae is scarce. From our knowledge, the only study focused on the technology and nutritional properties of GF bread supplemented with a cyanobacteria (*Arthrospira platensis*, known as Spirulina) was published by Figueira et al. [22], and more recently Rózylo and co-workers [23] determined the effect of brown macroalgae (*Ascophyllum nodosum*) on GF breads and observed increased antioxidant activity. In GF fresh pasta, the seaweed *Laminaria ochroleuca* showed promising potential to be used and similar mechanical and texture characteristics to the reference sample were achieved [12,13].

In the present study, a formulation based on buckwheat, rice flours and potato starch was enriched with *Tetraselmis chuii* biomass. Doughs prepared with the addition of microalgal biomass (1%, 2% and 4% *w/w*) were characterized using mixing tests in a MicrodoughLab and by oscillatory tests in a controlled-stress rheometer. The amount of water required to yield the same dough consistency of the previously optimized control-formulation was determined by the mixing test for the formulations enriched with microalgal biomass. Texture, volume and colour properties were used to evaluate the impact of different levels of *T. chuii* on the bread quality. Total phenolic content and reduction power of the bread extracts were assessed.

2. Materials and Methods

2.1. Raw Materials

Tetraselmis chuii was produced and collected by A2F partners in NORCE/UiB (Bergen, Norway). Cell wall disruption was applied to promote a controlled release of the active biocompounds, as it was recently proven in a previous study [18]. Fresh biomass was pre-processed by bead milling mechanical treatment in the pilot unit of NOFIMA and freeze-dried. It was determined to have high content of protein and also an important content of bioactive compounds: 47.7% protein, 11.4% lipids with 3.6% EPA, 2.4% starch and 2.3% salt [24].

Samples were prepared with the following flours: buckwheat flour (Próvida, Mem Martins, Portugal), rice flour (Espiga, Alcains, Portugal), potato starch (Globo, Seixal, Portugal), dried yeast (Fermipan®, Setúbal, Portugal), hydroxipropylmethylcellulose (HPMC) as a gelling agent (WellenceTM 321, Dow, Bomlitz, Germany), commercially available sugar, sunflower oil, salt, and water. The flours and HPMC were kindly supplied for free, except yeast, sugar, oil and salt that were purchased from local market. The commercial GF mix (Schär, Burgstall, Italy) has the following ingredients: corn starch, flax flour 12%, buckwheat flour 8%, pea bran, rice bran, apple fibre, sugar, guar seed flour, salt.

2.2. Preparation of the Samples

To compare the performance of GF breads with *T. chuii*, 2 controls were set up: a commercial GF mix to bake bread at home and a blank test without microalgae, called Control dough. The commercial GF bread was prepared from the mix, following the recommendations described on the product label.

GF breads with *T. chuii* (1.0 g, 2.0 and 4.0 g of *T. chuii*/100 g of flours + *T. chuii*) were prepared according to a previously optimised method [25]. This formulation, based on buckwheat and rice flours and potato starch, was tested using the specific volume and crumb firmness of the resulting breads as responses. Hydroxypropylmethylcellulose (HPMC), which is commonly used in GF breads, was used as a thickening agent, binding water and increasing doughs viscosity. The formulations studied are summarised in Table 1.

Table 1. Formulation of gluten-free (GF) samples and respective codes. Control: dough without microalgal biomass; Tc 1%, Tc 2%, Tc 4%: dough with 1%, 2% and 4% (*w/w*) *T. chuii*, respectively.

Ingredients (g/100g)	Control	Tc 1%	Tc 2%	Tc 4%
Buckwheat flour	46.0	45.5	45.1	44.2
Rice flour	31.0	30.7	30.4	29.8
Potato starch	23.0	22.8	22.5	22.1
Tetraselmis chuii	0.0	1.0	2.0	4.0
Sunflower oil (in relation to flours)	5.5	5.5	5.5	5.5
HPMC (in relation to flours)	4.6	4.6	4.6	4.6
Dried yeast (in relation to flours)	2.8	2.8	2.8	2.8
Sugar (in relation to flours)	2.8	2.8	2.8	2.8
Salt (in relation to flours)	1.8	1.8	1.8	1.8
Water Absorption (14% moisture basis)	69.0	69.0	69.0	69.0

Ingredients were mixed in a thermoprocessor equipment (Bimby—Vorwerk, Carnaxide, Portugal), initially to activate the yeast, by adding water, yeast and sugar for 2 min at 27 °C, at velocity 1. Following, the other ingredients were added and mixed for 10 min in a dough mixing program (wheat ear symbol). The resulted dough was placed in a fermentation chamber Arianna XLT133 (Unox, Cadoneghe, Italy) for 50 min at 37 °C. For breadmaking tests, the dough was baked in an electric oven Johnson A60 (Johnson & Johnson, New Brunswick, NJ, USA) at 180 °C for 50 min. Breads were analysed after cooling for 2 h. Three loaves of each formulation were prepared, and all the analysis were performed minimum in triplicate.

The pH values of the doughs ranged from 5.3 to 5.4, respectively for the commercial mix and control-dough, and increased with *T. chuii* incorporation (5.5 in Tc 1%, 5.7 in Tc 2% and 5.9 in Tc 4%).

2.3. Mixing Behaviour of the Dough

The Micro-doughLab 2800 (Perten Instruments, Sidney, Australia) was used to investigate the differences in formulation performances and determine the optimum water absorption capacity for each microalgal content (1% to 4% *w/w*). Tests were carried out using 4.00 ± 0.01 g mixed flours, at 14% moisture basis, using full formulation. The moisture of the different materials was measured through an automatic moisture analyser PMB 202 (Adam Equipment, Oxford, NJ, USA). Samples and water weights were corrected from flours and microalga moisture content. Standard manufacturer's protocol "General Flour Testing Method" was used, mixing the samples at a constant 63 rpm speed and temperature of 30 °C for 20 min. The peak value of torque of the optimised control-formulation was used, as a reference, to assess the optimum water absorption for each GF bread formulation with *T. chuii* addition. As a result, mixing curves and dough's mixing properties were assessed—peak resistance (mN.m), dough development time (s), stability (s) and softening (mN.m).

The amount of water added to the control dough (without microalgal biomass) was 69% on flours mixture basis. This water content was determined in the preliminary assays to produce breads having the best quality, based on bread volume and crumb firmness.

2.4. Viscoelastic Behaviour of the Dough

Small amplitude oscillatory shear (SAOS) measurements were performed in a controlled stress rheometer (Haake MARS III, Thermo Fisher Scientific, Waltham, MA, USA) equipped with a UTC-Peltier and fitted with a serrated parallel plate system with 20 mm diameter (PP20) and 1 mm gap (previously optimized for this type of material). After mixing, the dough was shaped into small balls and fermented in the oven at 37 °C, during 50 min. The fermented samples were placed between the plate sensor and the dough surface exposed was coated with paraffin oil to prevent drying, and allowed to rest at 5 °C ± 1 °C for 10 min before testing at same temperature to avoid fermentation during tests. Stress and frequency sweep tests were performed: stress sweep test at 6.28 rad/s (1 Hz) was always performed prior to the frequency sweep, to ensure testing within the linear viscoelastic zone. The viscoelastic properties of the dough were determined from the frequency sweep tests, applying a sinusoidal shear stress of 10 Pa (previously determined linear viscoelastic limit) over an angular frequency range of 0.0628 to 628 rad/s. For each sample, at least three repetitions were performed.

2.5. Evaluation of the Bread Texture

Bread texture was characterised using a Texturometer TA.XTplus (Stable Micro Systems, Surrey, UK) equipped with a 5 kg load cell, in a temperature controlled room at 20 ± 1 °C. Bread crumb was measured 2 h after baking by a puncture test using a cylindrical acrylic probe of 10 mm diameter (p/10) at 1 mm·s^{-1} crosshead speed and 10 mm penetration distance. Loaves were sliced by hand, with 20 mm thick. Measurements were repeated at least six times for each bread. Firmness (N) and cohesiveness were the texture parameters used to discriminate different bread samples [26].

2.6. Evaluation of the Bread Volume

Volume of the bread was measured using rapeseed displacement method AACC 10-05.01. In order to compare different breads, the same weight of ingredients, in relation to 300 g of flours in mixture with *T. chuii*, was used to prepare all the breads, using the formulations presented in Table 1. All the samples were evaluated, at least, in triplicate.

2.7. Evaluation of the Bread Colour

The bread colour was measured using a Minolta CR-400 (Japan) colorimeter with standard illuminant D65 and a visual angle of 2°. The results use the CIELAB system: L*—lightness (0 to 100), a*—greenness to redness (−60 to 60), and b*—blueness to yellowness (−60 to 60). The total colour difference between breads containing the microalgal biomass and the Control sample was calculated using the equation $\Delta E^* = [(\Delta L^*)^2 + (\Delta a^*)^2 + (\Delta b^*)^2]^{1/2}$. The measurements were replicated at least six times under artificial fluorescent light using a white standard (L* = 94.61, a* = −0.53, and b* = 3.62).

2.8. Evaluation of the Bread Bioactivity

To evaluate the bioactivity of breads, aqueous extracts were prepared by freeze-drying the bread samples and then milling into homogeneous fine powders using an electric blender. Then, 500 mg of each sample was stirred in 50 mL of distilled water using a magnetic bar for 30 min at room temperature, and then filtered by Whatman n° 4 paper. This procedure was repeated in duplicate for each bread. For all the bioactivity analysis, three replicates were performed for each extract, correspondent to six measurements for each bread sample.

The total phenolic content (TPC) of bread extract was evaluated using the method reported by Oktya et al. [27], with some modifications. The aqueous bread extract (300 µL) was added to 1500 µL of 0.1 mol/L *Folin–Ciocalteu* reagent and mixed with 1200 µL of sodium carbonate (7.5%) after 10 min. The mixtures were incubated in the dark at room temperature for 2 h, and then the absorbance was measured at 760 nm in a UNICAM UV4 UV/Vis Spectrometer. For blank testing, the extract was replaced by the same volume of water. The TPC was reported as milligrams of gallic acid equivalents per gram of extract (db).

The scavenging effect of bread extracts was determined using the DPPH (2,2-diphenyl-1-picryl-hydrazyl-hydrate) methodology [28]. Extraction solutions with a volume of 100 µL each were added to 1000 µL (90 µmol/L) of the DPPH solution in methanol, and the mixture was diluted with 1900 µL of methanol. After 1 h in the dark at room temperature, the absorbance was measured at 515 nm. In blank testing, the extract was replaced by the same volume of methanol. The reducing power of the bread extracts was determined using the ferric ion reducing antioxidant power (FRAP) assay [29]. The bread extract (90 µL) was added to 270 µL of water and 2.7 mL of FRAP reagent (2,4,6-tripyridyl-s-triazine (10 mmol/L) with HCl (40 mmol/L), FeCl3 (0.02 mol/L) and acetate buffer (0.3 mol/L pH 3.6) in a ratio of 1:1:10). After 30 min of incubation period in a 37 °C water bath (Thermo Scientific Precision 2864), the absorbance was measured at 593 nm wavelength. For the blank, the extract was substituted by the same volume of water. The mean values were reported as milligrams of ascorbic acid equivalents per gram of extract (db).

2.9. Sensory Evaluation

Hedonic sensory evaluation was performed for breads with 1 and 4% of T. chuii, as well as for control sample, using an untrained panel of 32 consumers, randomly chosen among the staff and students from the Instituto Superior de Agronomia Food Science Department, 14 males and 18 females, with ages between 16 and 65. The three breads were analysed in terms of colour, smell, taste, texture and general acceptance using a 5-point hedonic scale from "very unpleasant" (1) to "very pleasant" (5). The assays were conducted in a standardized sensory analysis room, according to the standard EN ISO 8589: 2007.

2.10. Statistical Analysis

The analysis of variance (one-way ANOVA) of the experimental data was performed using Origin Pro 8.0 software (OriginLab Corporation, Northampton, MA, USA), followed by Tukey's test. Correlation analysis was performed by using STATISTICA (version 10.0, StatSoft, Hamburg, Germany) and the function Bivariate Scatterplot. The significance level was set to 95% ($p < 0.05$).

3. Results and Discussion

3.1. Impact of Tetraselmis chuii Addition on the Dough Rheology

3.1.1. Empirical Methods—MicrodoughLab

The mixing curves obtained from the MicrodoughLab are represented in Figure 1, where resistance to mixing is measured as torque. The manufacturer's protocol "General Flour Testing Method", at slow speed, was adopted. GF bread doughs have a completely different composition and structure from the wheat dough. These GF systems are mainly composed of starch granules embedded in a matrix composed of proteins and other hydrocolloids, but without a gluten network [21]. GF doughs are adhesive and have poor mixing properties. The mixing curves showed a high initial torque as water was hydrating the flours, followed by torque decrease and unstable mixing curves were obtained, without a peak development as observed in wheat doughs mixing curves. Although this test was developed for wheat dough, and aims a targeted peak of 100 mN.m torque, for an optimum consistency, it is not possible to use this value in GF doughs. Therefore, different approaches have been used to determine the water absorption level of GF formulations. Some researchers carried out preliminary assays to optimise the amount of water necessary for bread making process based on the specific volume of the produced bread [30,31] and crumb hardness of the resulting breads [30]. Others, used the farinograph mixer in order to achieve an optimal dough consistency of 200 BU [32] or 500 BU—Brabender units [33], dough consistency checked by the texturometer equipped with back extrusion rig, using a reference value of firmness [34], comparison of complex viscosity (η^*) values vs. frequency, taking a soft wheat flour dough which gave a peak at 500 BU as a reference [35], etc.

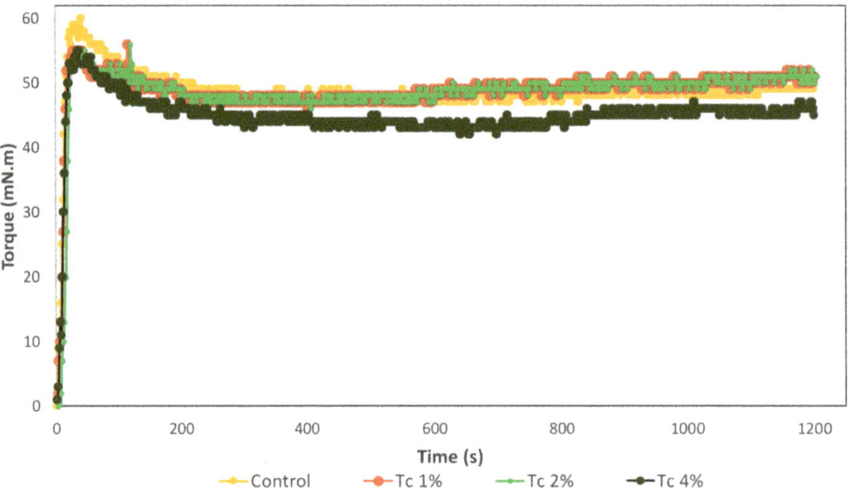

Figure 1. Mixing curves obtained from the Micro-doughLab of Control and GF doughs with *T. chuii* Tc 1%, Tc 2% and Tc 4%.

Our proposed methodology is to assess optimum water absorption values (amount of water needed to achieve target peak torque, corrected to 14% moisture basis) for the formulations containing microalgal biomass in order to achieve a peak of 56 nN.m ± 5 mN.m. This peak value corresponds to a control-dough (same GF formulation without microalgae) with good bread baking characteristics, previously optimised by testing different water hydrations (data not presented). The technological parameters obtained from the mixing curves are presented in Table 2. For the microalgae contents considered in the present study (1 to 4% *w/w*) and 69% of water absorption, similar peak values were obtained, without being necessary to adjust the water content of microalgal-containing formulations. From this empirical rheology test, it was concluded that *T. chuii* addition had no significant ($p < 0.05$) impact on the rheology parameters extracted from de mixing curves: optimum water absorption, peak development time, stability and softening.

Table 2. Parameters obtained from Micro-doughLab mixture curves of Control and GF dough with *T. chuii* Tc 1%, Tc 2% and Tc 4%.

GF Dough	WA (%)	P (mN.m)	DDT (s)	DS (s)	DSO (mN.m)
Control	69.0	56 ± 1.2	48 a	30 a	9 a
Tc 1%	69.0	51 ± 1.7	48 a	44 a	7 a
Tc 2%	69.0	56 ± 3.8	50 a	44 a	7 a
Tc 4%	69.0	51 ± 1.5	48 a	50 a	9 a

Note: WA—water absorption; P—peak resistance; DDT—dough development time; DS—dough stability; DSO—dough softening; PE—peak energy. For each sample, the test was performed in triplicate. Different letters in the same column correspond to significant differences ($p < 0.05$).

The GF mix was not evaluated using Micro-doughLab since water hydration followed the recipe indicated on the label of this commercial product.

3.1.2. Fundamental Methods—Small Amplitude Oscillatory Shear Measurements (SAOS)

Frequency sweeps were performed to evaluate the impact of *T. chuii* biomass addition on dough structure, after fermentation (Figure 2). For each sample, the test was performed in triplicate, and the most representative curve for each sample is presented. Doughs have a viscoelastic behaviour with G′ > G″, for the whole range of frequencies studied, showing a destructuring effect, resulting from the microalgae addition, that was observed at 1% and 2% levels. However, for higher Tc incorporation, 4%, an increase of G′ values was observed. For the 4% *T. chuii* dough, the values of the elastic modulus (G´ at 6.283 rad/s and 62.83 rad/s) are significantly ($p < 0.05$) higher than for 2% *T. chuii*, and are similar to the commercial mix, control-sample and 1% *T. chuii*. These values correspond to a structure recovery at 4% level of microalgae addition.

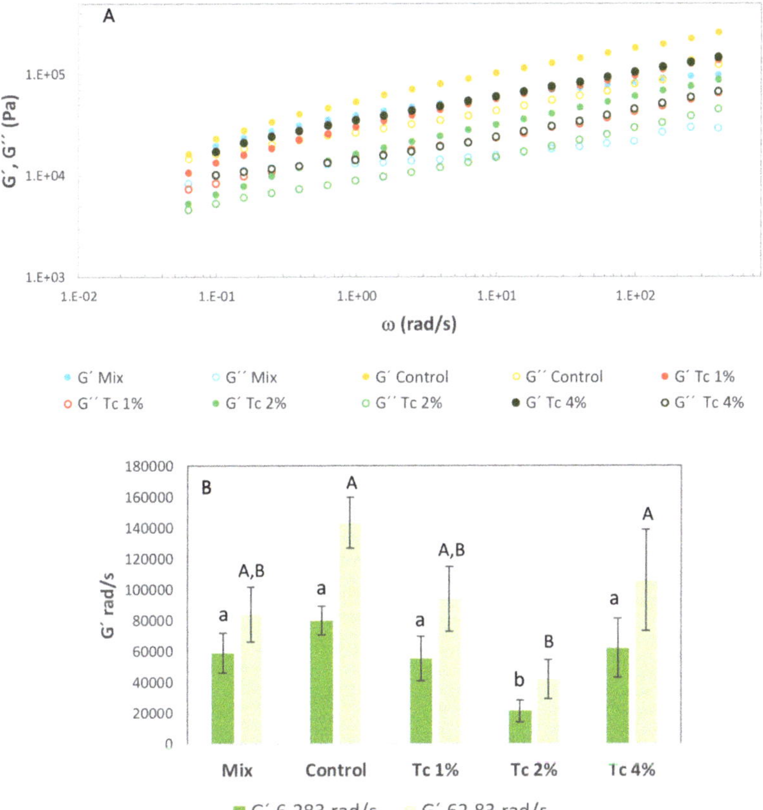

Figure 2. (**A**) Mechanical spectra and (**B**) values of G′ at 6.283 rad/s (1 Hz) and 62.83 rad/s (10 Hz) obtained after GF dough fermentation. G′ (storage modulus—filled symbol), G″ (loss modulus—open symbol). Mix, Control and GF dough with *T. chuii* Tc 1%, Tc 2% and Tc 4%. Error bars indicate the standard deviations from the repetitions. Different letters correspond to significant differences ($p < 0.05$).

The frequency dependence of G′ (elastic modulus) and G″ (viscous modulus) could be described by the power law Equations (1) and (2):

$$G' = a' \, \omega^{b'} \tag{1}$$

$$G'' = a'' \, \omega^{b''} \tag{2}$$

Values of a and b are determined by performing a linear regression on log G′ and G″ versus log frequency, where a′ and a″ are the y-intercepts and b′ and b″ are the slopes of the resulting line [7]. According to a′ and a″ values, Tc 2% dough presented a lower level of structure, showing the lower a′ value and the highest value of b′ (Table 3).

Table 3. Frequency dependence of G′ and G″ described by the power-law parameters. Mix, Control and GF dough with T. chuii Tc 1%, Tc 2% and Tc 4%. Different letters in the same column correspond to significant differences ($p < 0.05$). $R^2 > 0.91$.

GF Dough	a′	b′	a″	b″
Mix	38397 [a]	0.182 [d]	13942 [a,b]	0.130 [a]
Control	43446 [a]	0.289 [a,b]	23896 [a]	0.239 [a]
Tc 1%	31724 [a,b]	0.267 [b,c]	15642 [a,b]	1.497 [a]
Tc 2%	11639 [b]	0.318 [a]	6875 [b]	0.268 [a]
Tc 4%	37213 [a]	0.249 [c]	17555 [a]	0.223 [a]

Flours from grains without gluten, such as rice, and from pseudocereals, such as buckwheat, have been used as ingredients to produce GF breads. Addition of pseudocereals considerably improves dough viscosity and the texture, volume and nutritional quality of GF products [20,36] and could lead to improved anti-staling properties [37]. To our knowledge, studies relating the impact of microalgae in GF dough rheology have not been published so far. For wheat bread, Nunes et al. [18] have studied the impact of microalgae cell disruption pretreatment on the dough rheology and bread bioactivity and found that 1% *Chlorella vulgaris* has a negative impact on the dynamic viscoelastic properties of the wheat dough.

The structure of complex GF dough systems is mainly accounted by starch and HPMC which could entrap the air. When small amounts of protein are added, coming from *T. chuii*, a destabilisation of this structure can be noticed (Figure 2), mainly for the sample with 2% of microalgal incorporation, however there is a structure recovery for 4% addition. It looks like microalgae elements at first disrupt the structure formed by starch granules embedded in amylose matrix reinforced by HPMC long rods, this can probably happen by depletion flocculation by antagonism with protein macromolecules from *T. chuii*, up to a certain concentration. For higher levels (4%), the proteins should take the lead in the structure network, playing an important role by replacing the former structure, contributing to a new matrix, dominated by protein interactions, where the starch granules and HPMC will be embedded with some compatible reinforcement of the amylose released from starch granules. This structure is different from the previous one, more compact, as it can be seen by the smaller size of the crumb alveoli.

3.2. Impact of Tetraselmis chuii Addition on the Breadmaking Properties

The impact of *T. chuii* incorporation in the GF bread shape and colour can be accessed through Figure 3.

Figure 3. General aspect of the GF breads prepared with commercial mix, control and 1, 2 and 4% (*w/w*) *T. chuii* gluten free breads.

To study the relation between overall bread quality parameters, raw materials and dough properties, the texture and the volume of bread loaves were evaluated. For bread crumb differing in composition, crumb firmness increased with microalgal addition until 2% *T. chuii* (Figure 4) and there was a negative impact on loaf volume (Table 4). However, at 4% level, there was an increase of the bread volume and a significant ($p < 0.05$) reduction of bread crumb firmness comparing to 2%, not differing from 1% *T. chuii* bread. Nevertheless, this formulation with 4% of microalgal biomass presented a significant ($p < 0.05$) lower cohesiveness when compared to all the others, including the control-bread and the bread prepared with a commercial mix.

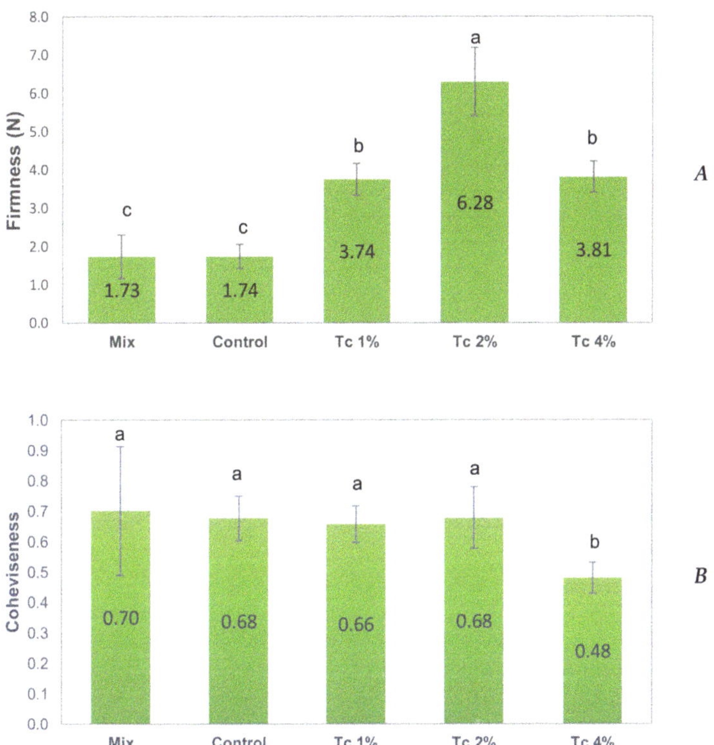

Figure 4. Firmness (**A**) and cohesiveness (**B**) values of GF doughs with *T. chuii* obtained by the texturometer. Mix, Control and GF dough with *T. chuii* Tc 1%, Tc 2% and Tc 4%. Error bars indicate the standard deviations from the repetitions ($n = 6$). Different letters in the same graph correspond to significant differences ($p < 0.05$).

Table 4. Bread volume and colour parameters obtained for bread crumb. Mix, Control and GF dough with *T. chuii* Tc 1%, Tc 2% and Tc 4%. Different letters in the same column correspond to significant differences ($p < 0.05$).

GF Bread	Volume (cm³)	L*	a*	b*	ΔE*
Mix	674 [b]	89.3 [a]	8.5 [a]	24.0 [a,b]	–
Control	701 [a]	94.6 [a]	1.4 [b]	19.2 [c]	–
Tc 1%	642 [c]	38.9 [b]	−1.1 [c]	16.2 [d]	64
Tc 2%	612 [d]	29.8 [b,c]	−1.3 [c]	27.1 [a]	73
Tc 4%	640 [c]	25.7 [c]	−0.3 [c]	22.2 [b,c]	77

Starch gelatinization plays an important role in GF formulations, due to the ability of starch to form a matrix in which gas bubbles are entrapped [38]. Incorporation of non-gluten proteins in the formulation, even at lower amounts, may be essential in achieving final product volume [30,39,40].

Considering the present results, it is possible to conclude that at low contents of *T. chuii* incorporation, microalgal protein induces a destabilisation of the network formed by starch and HPMC. This is revelled by a significant reduction of bread volume (the bread becomes more compact), as well as a significant increase of firmness. However, for the highest protein content of 4%, microalgal protein plays an important role on the GF dough structure, increasing the capacity to retain the gas bubbles.

In what crumb colour is concerned (Table 4), it is possible to conclude that there was a reduction in lightness (L*) and an increase in greeness (a*, in modulus), comparing to commercial mix and control breads. In respect to yellowness (b*), there was a significant ($p < 0.05$) decrease from control (19.2) to 1% *T. chuii* bread (16.2), increasing with 2% of algae (27.1) and decreasing again for 4% *T. chuii* (22.2). These results are related to the pigments of this *Chlorophyceae* green algae, namely chlorophylls, carotenoids (such as fucoxanthin and β-carotene) and α-tocopherol (Vitamin E) [41]. The total colour difference ΔE* between control and *T. chuii* crumbs was higher than 64, despite the small differences between 2% and 4% *T. chuii* bread colours. Several authors consider that the human eye can differentiate colours when the total colour difference ΔE* > 5 [42], therefore big colour changes were obtained when breads were enriched with microalgal biomass. From these results, it seems that *T. chuii* is contributing to a darkening effect, as it is evident by visual observation (Figure 3) and through the instrumental colour parameters. As will be described in Section 3.5., these colour characteristics are not well appreciated by the sensory panel.

3.3. Correlations between Breadmaking Properties and Dough Rheology

Using correlation analysis, the relationships between dough rheology and breadmaking performance of breads with different amounts of *T. chuii* incorporation were obtained. In Figure 5, one can see the bivariate scatterplots of significant ($r > 0.70$) dependences - bread firmness vs elastic modulus ($G'_{6.283\ rad/s}$), bread volume vs elastic modulus ($G'_{6.283\ rad/s}$), and bread volume vs bread firmness. High G' values are related with greater elastic contribution and this is negatively correlated with the bread firmness (softening effect), followed by a bread volume increase. As expected, bread volume presented a negative correlation with bread firmness.

Similar results were described by Martínez & Gómez [43], showing that viscoelastic properties of several GF doughs strongly influenced the bread volume and crumb texture. Elgeti et al. [44] referred that the rheology of starch-based dough systems influences the level of aeration during mixing and baking.

3.4. Impact of Tetraselmis chuii Addition on the Bread Bioactivity

Total phenolic content and antioxidant capacity (DPPH and FRAP) were evaluated only for the control and 4% *T. chuii* breads, considering the previous results and the objective of achieving high levels of microalgae incorporation.

The addition of the microalgal biomass at 4% (*w/w*) resulted in a significant ($p < 0.05$) increase in the total phenolic content (TPC) (Figure 6), which was 0.11 mg·g^{-1} gallic acid equivalents in the Control bread and 0.24 mg·g^{-1} in the bread containing *T. chuii*. TPC values were lower than the values reported for GF breads with brown microalgae [23] and may be a result of the formation of protein-phenolic complexes [23,45].

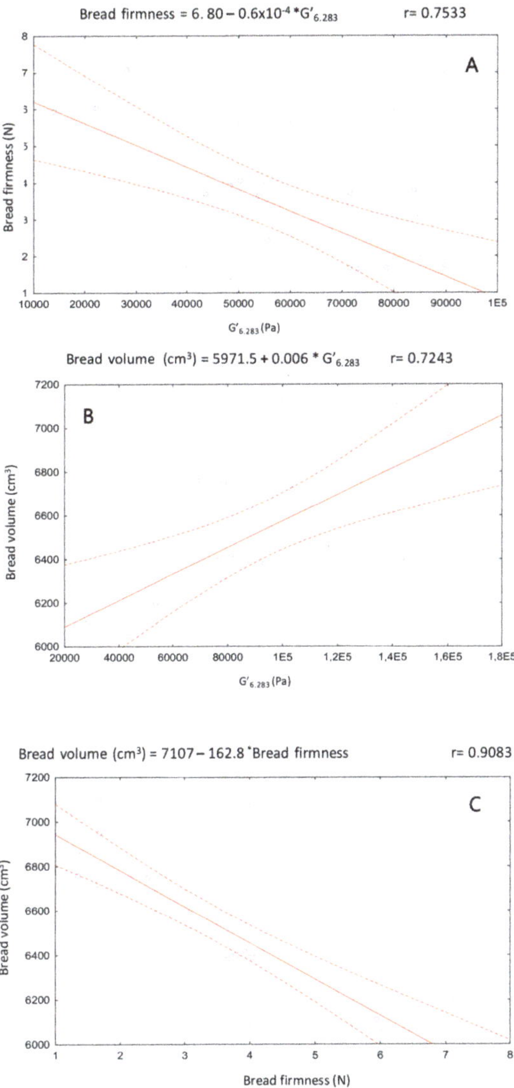

Figure 5. Mathematical correlations between GF breadmaking properties and dough rheology ($p < 0.05$). (**A**) Bread firmness and $G'_{6.283\ \text{rad/s}}$; (**B**) Bread volume and $G'_{6.263\ \text{rad/s}}$; (**C**) Bread volume and bread firmness.

The antioxidant capacity was tested using the DPPH and FRAP methods. Compared with the control-bread (2.75 mg·g^{-1} and 0.33 mg·g^{-1} ascorbic acid equivalents obtained by DPPH and FRAP, respectively), the incorporation of microalgal biomass led to a significant ($p < 0.05$) increase in the antioxidant capacity of the breads (3.22 mg·g^{-1} and 0.47 mg·g^{-1} ascorbic acid equivalents). Even upon baking, the antioxidant activity of these breads is interesting. Some other authors reported similar results, Rózylo et al. [23] determined the effect of brown macroalgae on GF bread and observed increased FRAP activity, and higher FRAP antioxidant capacity was found by Nunes et al. [18] in wheat breads enriched with *Chlorella vulgaris*.

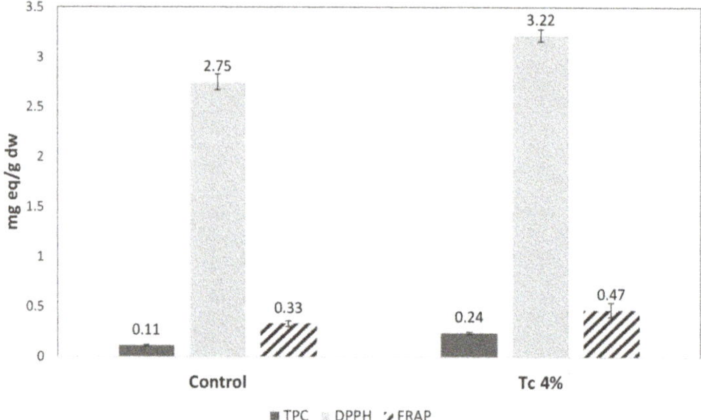

Figure 6. Total phenolic content (mg·g^{-1} gallic acid equivalents), antioxidant capacity measured using the DPPH assay (2,2-diphenyl-1-picryl-hydrazyl-hydrate; mg.g^{-1} ascorbic acid equivalents) and FRAP assay (ferric ion reducing antioxidant power; mg.g^{-1} ascorbic acid equivalents) of GF breads enriched with 4% (*w/w*) of *T. chuii* biomass in comparison with Control bread. Error bars indicate the standard deviations of the repetitions (n = 3).

3.5. Sensory Evaluation

Sensory analysis assays were carried out with *T. chuii* microalgal GF breads, at 1% and 4% incorporation level, in comparison to control. Figure 7 represents the average scores of the sensory parameters as evaluated by the non-celiac panel.

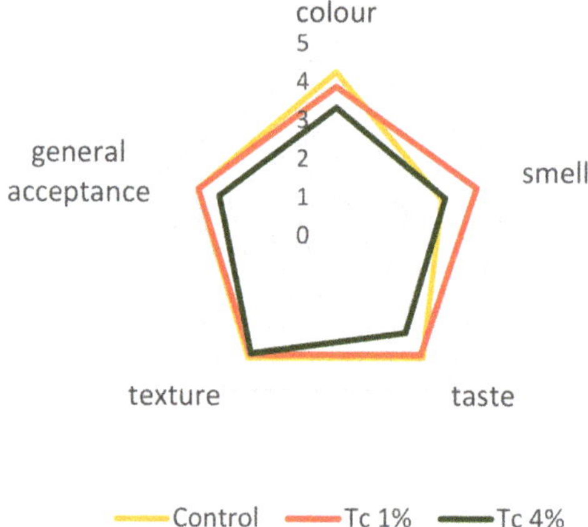

Figure 7. Responses of the sensory analysis panel tasters (*n* = 32) regarding breads enriched with 1% and 4% *T. chuii*, as well as the control sample. 1—"very unpleasant", 2—"unpleasant", 3—"indifferent", 4—"pleasant", 5—"very pleasant".

Panellists preferred control bread and bread with 1% *T. chuii*, both with a general acceptance of 3.84, compared to 3.25 of 4% *T. chuii* bread. Texture, colour and taste had similar score for control and 1% *T. chuii* bread. Concerning smell, the tasters preferred the bread with 1% *T. chuii* in relation to control, by a difference of 1.0 point. This can be related to the buckwheat flour, despite the characteristic fishy flavour of *T. chuii* microalgae. Eventhough, for higher microalgal levels, 4% *T. chuii* and control bread had similar scores. As expected for this type of product, for taste, colour and general acceptance, bread with 4% *T. chuii* had lower scores, but not lower than 3, corresponding to "indifferent". In the comments field of the sensory analysis sheet, some tastes referred the strong fishy flavour of the 4% *T. chuii* bread, but that it could be pleasant to eat with fish meals. Furthermore, this bread could be an interesting alternative for consumers interested in healthy products with an innovative taste and colour. To improve the acceptance by the conventional consumers, educational marketing strategies and formulation enhancements are programmed in the Algae2Future project.

4. Conclusions

The mixing and viscoelastic behaviour of GF doughs enriched with *Tetraselmis chuii* was compared with the control formulation. Bread baking performance was also evaluated since GF doughs are complex systems and final bread quality is affected by processing conditions.

T. chuii can be used as a natural novel and sustainable ingredient to increase the bioactivity of GF bread based on buckwheat flour, rice flour and potato starch, obtaining an innovative green appearance. Different behaviour was found according with the level of *T. chuii* incorporation. Below 2%, *T. chuii* proteins destabilize the structure developed by starch and HPMC, smaller bread volume was obtained, associated with a more compact crumb and harder properties. However, for higher levels of incorporation (4%), the microalgal proteins with starch and HPMC build up another type of structure, characterised by higher values of the viscoelastic functions (G'and G") producing higher bread volume and a softening effect. This study shows that the structure of 4% *T. chuii* bread is competitive with the control-bread with the advantage of having an improved bioactivity (phenolics and antioxidants) with possible positive impacts on health. The use of *T. chuii* at 4% level is interesting, but low sensory scores postpone its utilization.

Author Contributions: M.C.N., I.S. and A.R. oversaw conception and design. M.C.N. and I.F. oversaw bread formulation and preparation, physical analysis and data interpretation. M.C.N. and I.V. oversaw bread bioactivity analysis and data interpretation. M.C.N. was responsible for statistical analysis and redaction of the manuscript. M.C.N., I.S. and A.R. were responsible for the discussion of the results and revision of the manuscript. All authors have read and agreed to the published version of the manuscript.

Funding: Research Council of Norway, Algae to Future (A2F) Project - From Fundamental Algae Research to Applied Industrial Practices. Portuguese Foundation for Science and Technology (FCT), UID/AGR/04129/2013—LEAF.

Acknowledgments: The authors wish to thank Dorinde Kleinegris from NORCE/UiB in Bergen (Norway) for *Tetraselmis chuii* microalgal biomass production at pilot scale, as well as Maria Eleni Kokkali and Katerina Kousoulaki from NOFIMA in Bergen (Norway) for cell wall disruption and freeze drying of *T. chuii* biomass.

Conflicts of Interest: None of the authors has potential financial or other conflicts of interest to disclose.

References

1. Koyande, A.K.; Chew, K.W.; Rambabu, K.; Tao, Y.; Chu, D.-T.; Show, P.-L. Microalgae: A potential alternative to health supplementation for humans. *Food Sci. Hum. Wellness* **2019**, *8*, 16–24. [CrossRef]
2. Leach, A.M.L.; Gallowaya, J.N.; Bleeker, A.; Erisman, J.W.; Khon, R.; Kitzes, J. A nitrogen footprint model to help consumers understand their role in nitrogen losses to the environment. *Environ. Dev.* **2012**, *1*, 40–66. [CrossRef]
3. Batista, A.P.; Gouveia, L.; Bandarra, N.M.; Franco, J.M.; Raymundo, A. Comparison of microalgal biomass profiles as novel functional ingredient for food products. *Algal Res.* **2013**, *2*, 164–173. [CrossRef]
4. Khan, M.I.; Shin, J.H.; Kim, J.D. The promising future of microalgae: Current status, challenges, and optimization of a sustainable and renewable industry for biofuels, feed, and other products. *Microb. Cell Factories* **2018**, *17*, 36. [CrossRef]
5. Lafarga, T. Effect of microalgal biomass incorporation into foods: Nutritional and sensorial attributes of the end products. *Algal Res.* **2019**, *41*, 101566. [CrossRef]
6. Lafarga, T.; Mayre, E.; Echeverria, G.; Viñas, I.; Villaró, S.; Acién-Fernández, F.G.; Castellari, M.; Aguiló-Aguayo, I. Potential of the microalgae *Nannochloropsis* and *Tetraselmis* for being used as innovative ingredients in baked goods. *LWT Food Sci. Technol.* **2019**, *115*, 108439. [CrossRef]

7. Bernaerts, T.M.M.; Panozzo, A.; Doumen, V.; Foubert, I.; Gheysen, L.; Goiris, K.; Moldenaers, P.; Hendrickx, M.E.; Van Loey, A.M. Microalgal biomass as a (multi)functional ingredient in food products: Rheological properties of microalgal suspensions as affected by mechanical and thermal processing. *Algal Res.* **2017**, *25*, 452–463. [CrossRef]

8. Bernaerts, T.M.M.; Gheysen, L.; Foubert, I.; Hendrickx, M.E.; Van Loey, A.M. The potential of microalgae and their biopolymers as structuring ingredients in food: A review. *Biotechnol. Adv.* **2019**, *37*, 107419. [CrossRef] [PubMed]

9. Raymundo, A.; Nunes, M.C.; Sousa, I. Microalgae biomass as a food ingredient to design added value products. In *Valorising Seaweed by-Products*; Domínguez, H., Torres, M.D., Eds.; Nova Science Publishers: Hauppauge, NY, USA, 2019; ISBN 978-1-53615-398-9.

10. Fradique, M.; Batista, A.P.; Nunes, M.C.; Gouveia, L.; Bandarra, N.M.; Raymundo, A. Incorporation of *Chlorella vulgaris* and *Spirulina maxima* biomass in pasta products. Part I: Preparation and evaluation. *J. Sci. Food Agric.* **2010**, *90*, 1656–1664. [CrossRef] [PubMed]

11. Fradique, M.; Batista, A.P.; Nunes, M.C.; Gouveia, L.; Bandarra, N.; Raymundo, A. *Isochrysis galbana* and *Diacronema vlkianum* biomass incorporation in pasta products as PUFA's source. *LWT Food Sci. Technol.* **2013**, *50*, 312–319. [CrossRef]

12. Fradinho, P.; Raymundo, A.; Sousa, I.; Domínguez, H.; Torres, M.D. Edible brown seaweed in gluten-free pasta: Technological and nutritional evaluation. *Foods* **2019**, *8*, 622. [CrossRef] [PubMed]

13. Fradinho, P.; Niccolai, A.; Soares, R.; Rodolfi, L.; Biondi, N.; Tredici, M.R.; Sousa, I.; Raymundo, A. Effect of *Arthrospira platensis* (spirulina) incorporation on the rheological and bioactive properties of gluten-free fresh pasta. *Algal Res.* **2020**, *45*, 101743. [CrossRef]

14. Gouveia, L.; Coutinho, C.; Mendonça, E.; Batista, A.P.; Sousa, I.; Bandarra, N.M.; Raymundo, A. Functional biscuits with PUFA-ω3 from *Isochrysis galbana*. *J. Sci. Food Agric.* **2008**, *88*, 891–896. [CrossRef]

15. Batista, A.P.; Nicccolai, A.; Fradinho, P.; Fragoso, S.; Bursic, I.; Rodolfi, L.; Biondi, N.; Tredici, M.R.; Sousa, I.; Raymundo, A. Microalgae biomass as an alternative ingredient in cookies: Sensory, physical and chemical properties, antioxidant activity and in vitro digestibility. *Algal Res.* **2017**, *26*, 161–171. [CrossRef]

16. Batista, A.P.; Niccolai, A.; Bursic, I.; Sousa, I.; Raymundo, A.; Rodolfi, L.; Biondi, N.; Tredici, M.R. Microalgae as funtional ingredientes in savory food products: Application to wheat crackers. *Foods* **2019**, *8*, 611. [CrossRef]

17. Graça, C.; Fradinho, P.; Sousa, I.; Raymundo, A. Impact of *Clorella vulgaris* addition on rheology wheat dough properties. *LWT Food Sci. Technol.* **2018**, *89*, 466–474. [CrossRef]

18. Nunes, M.C.; Graça, C.; Vlaisavljevic, S.; Tenreiro, A.; Sousa, I.; Raymundo, A. Microalgae cell disruption: Effect on the bioactivity and rheology of wheat bread. *Algal Res.* **2020**, *45*, 101749. [CrossRef]

19. Matos, M.E.; Rosell, C.M. Understanding gluten-free dough for reaching breads with physical quality and nutritional balance. *J. Sci. Food Agric.* **2015**, *95*, 653–661. [CrossRef]

20. Naqash, F.; Gani, A.; Gani, A.; Masoodi, F.A. Gluten-free baking: Combating the challenges – a review. *Trends Food Sci. Technol.* **2017**, *66*, 98–107. [CrossRef]

21. El Khoury, D.; Balfour-Ducharme, S.; Joye, I.J. A review on the gluten-free diet: Technological and nutritional challenges. *Nutrients* **2018**, *10*, 1410. [CrossRef]

22. Figueira, F.S.; Crizel, T.M.; Silva, C.R.; Salas-Mellado, M.M. Elaboration of gluten-free bread enriched with the microalgae *Spirulina platensis*. *Braz. J. Food Technol.* **2011**, *14*, 308–316.

23. Rózylo, R.; Hassoon, W.H.; Gawlik-Dziki, U.; Siastala, M.; Dziki, D. Study on the physical and antioxidant properties of gluten-free bread with brown algae. *CyTA J. Food* **2017**, *15*, 196–203. [CrossRef]

24. Kokkali, M.E.; Kleinegris, D.M.M.; de Vree, J.H.; Haugsgjerd, B.O.; Samuelsen, T.A.; Oterhals, A.; Kousoulaki, K. Optimising cell wall disruption by bead milling of microalgae biomass for release of nutrients for aquafeed and food applications (preliminary results). In Proceedings of the HydroMedit 2018, 3rd International Congress on Applied Ichthyology & Aquatic Environment, Volos, Greece, 8–11 November 2018.

25. Fernandes, I.C.X. Desenvolvimento de Pães Sem Glúten Enriquecidos com *Tetraselmis chuii*. Master's Thesis, Instituto Superior de Agronomia da Universidade de Lisboa, Lisboa, Portugal, 2019.

26. Bourne, M. *Food Texture and Viscosity*, 2nd ed.; Academic Press: London, UK, 2002.

27. Oktay, M.; Gülçin, İ.; Kufrevioglu, O.İ. Determination of in vitro antioxidant activity of fennel (Foeniculum vulgare) seed extracts. *LWT Food Sci. Technol.* **2003**, *36*, 263–271. [CrossRef]

28. Sánchez-Moreno, C.; Larrauri, J.A.; Saura-Calixto, F. A procedure to measure the antiradical efficiency of polyphenols. *J. Sci. Food Agric.* **1998**, *76*, 270–276. [CrossRef]
29. Rufino, M.S.M.; Alves, R.E.; Brito, E.S.; Morais, S.M.; Sampaio, C.G.; Pérez-Jiménez, J.; Saura-Calixto, F.D. *Metodologia Científica: Determinação da Atividade Antioxidante Total em Frutas pelo Método de Redução do Ferro (FRAP)*; Embrapa Agroindústria Tropical: Fortaleza Brasil, 2006.
30. Storck, C.R.; Zavareze, E.R.; Gularte, M.A.; Elias, M.C.; Rosell, C.M.; Dias, A.R.G. Protein enrichment and its effects on gluten-free bread characteristics. *LWT Food Sci. Technol.* **2013**, *53*, 346–354. [CrossRef]
31. Sahagún, M.; Gómez, M. Assessing influence of protein source on characteristics of gluten-free breads optimising their hydration level. *Food Bioprocess Technol.* **2018**, *11*, 1686–1694. [CrossRef]
32. Mariotti, M.; Lucisano, M.; Pagani, M.A.; Ng, P.K.W. The role of corn starch, amaranth flour, pea isolate, and Psyllium flour on the rheological properties and the ultrastructure of gluten-free doughs. *Food Res. Int.* **2009**, *42*, 963–975. [CrossRef]
33. Mohammadi, M.; Azizi, M.H.; Neyestani, T.R.; Hosseini, H.; Mortazavian, A.M. Development of gluten-free bread using guar gum and transglutaminase. *Ind. Eng. Chem.* **2015**, *21*, 1398–1402. [CrossRef]
34. Korus, J.; Witczak, M.; Ziobro, R.; Juszczak, L. The influence of acorn flour on rheological properties of gluten-free dough and physical characteristics of the bread. *Eur. Food Res. Technol.* **2015**, *240*, 1135–1143. [CrossRef]
35. Yazar, G.; Duvarci, O.; Tavman, S.; Kokini, J.L. Non-linear rheological behaviour of gluten-free flour doughs and correlations of LAOS parameters with gluten-free bread properties. *J. Cereal Sci.* **2017**, *74*, 28–36. [CrossRef]
36. Wang, K.; Lu, F.; Li, Z.; Zhao, L.; Han, C. Recent developments in gluten-free bread baking approaches: A review. *Food Sci. Technol* **2017**, *37*, 1–9. [CrossRef]
37. Torbica, A.; Hadnacev, M.; Dapcevic, T. Rheological, textural and sensory properties of gluten-free bread formulations based on rice and buckwheat flour. *Food Hydrocoll.* **2010**, *24*, 626–632. [CrossRef]
38. Horstmann, S.W.; Lynch, K.M.; Arendt, E.K. Starch characteristics linked to gluten-free products. *Foods* **2017**, *6*, 29. [CrossRef] [PubMed]
39. Graça, C.; Raymundo, A.; Sousa, I. Yogurt as an alternative ingredient to improve the functional and nutritional properties of gluten-free breads. *Foods* **2020**, *9*, 111. [CrossRef]
40. Tomic, J.; Torbica, A.; Belovic, M. Effect of non-gluten proteins and transglutaminase on dough rheological properties and quality of bread based on millet (*Panicum miliaceum*) flour. *LWT Food Sci. Technol.* **2020**, *118*, 108852. [CrossRef]
41. Carballo-Cárdenas, E.C.; Tuan, P.M.; Janssen, M.; Wijffels, R.H. Vitamin E (α-tocopherol) production by marine microalgae *Dunaliella tertiolecta* and *Tetraselmis suecica* in batch cultivation. *Biomol. Eng.* **2003**, *20*, 139–147. [CrossRef]
42. Francis, F.J.; Clydesdale, F.M. *Food Colorimetry: Theory and Applications*; The AVI Publishing Company Inc.: Westport, CT, USA, 1975.
43. Martínez, M.M.; Gómez, M. Rheological and microstructural evolution of the most common gluten-free flours and starches during bread fermentation and baking. *J Food Eng.* **2017**, *197*, 78–86. [CrossRef]
44. Elgeti, D.; Peng, L.; Jekle, M.; Becker, T. Foam stabilization during processing of starch-based dough systems. *Innov. Food Sci. Emerg. Technol.* **2017**, *39*, 267–274. [CrossRef]
45. Ballester-Sánchez, J.; Gil, J.V.; Haros, C.M.; Fernández-Espinar, M.T. Effect of Incorporating white, red or black quinoa flours on free and bound polyphenol content, antioxidant activity and colour of bread. *Plant Foods Hum. Nutr.* **2019**, *74*, 185–191. [CrossRef]

Article

Impact of Acorn Flour on Gluten-Free Dough Rheology Properties

R. Beltrão Martins [1,2,*], M. C. Nunes [3], L. M. M. Ferreira [1], J. A. Peres [2], A. I. R. N. A. Barros [1] and A. Raymundo [3]

[1] CITAB—Centre for the Research and Technology of Agro-Environmental and Biological Sciences, Universidade de Trás-os-Montes e Alto Douro, 5000-801 Vila Real, Portugal; lmf@utad.pt (L.M.M.F.); abarros@utad.pt (A.I.R.N.A.B.)
[2] CQVR—Chemistry Research Centre, Chemistry Department, Universidade de Trás-os-Montes e Alto Douro, 5000-801 Vila Real, Portugal; jperes@utad.pt
[3] LEAF—Linking Landscape, Environment, Agriculture and Food, Instituto Superior de Agronomia, Universidade de Lisboa, Tapada da Ajuda, 1349-017 Lisbon, Portugal; crnunes@gmail.com (M.C.N.); anabraymundo@isa.ulisboa.pt (A.R.)
* Correspondence: ritabeltraomartins@icloud.com; Tel.: +351-966-908-963

Received: 31 March 2020; Accepted: 26 April 2020; Published: 2 May 2020

Abstract: Gluten is a fundamental ingredient in breadmaking, since is responsible for the viscoelastic behaviour of the dough. The lack of gluten has a critical effect on gluten-free dough, leading to less cohesive and less elastic doughs, and its replacement represents a challenge for bakery industry. However, dough rheology can be improved combining different ingredients with structural capacity and taking advantage from their interactions. Although acorn flour was used to bake bread even before Romans, nowadays is an underexploited resource. It presents good nutritional characteristics, particularly high fibre content and is naturally gluten free. The aim of this study was to use acorn flour as a gluten-free ingredient to improve dough rheology, following also market trends of sustainability and fibre-rich ingredients. Doughs were prepared with buckwheat and rice flours, potato starch and hydroxypropylmethylcellulose. Two levels of acorn flour (23% and 35% *w/w*) were tested and compared with control formulation. Micro-doughLAB was used to study mixing and pasting properties. Doughs were characterised using small amplitude oscillatory measurements (SAOS), with a controlled stress rheometer, and regarding Texture Profile Analysis (TPA) by a texturometer. Dietary fibre content and its soluble and insoluble fractions were also evaluated on the developed breads. Acorn flour showed promising technological properties as food ingredient for gluten-free baking (improved firmness, cohesiveness and viscoelasticity of the fermented dough), being an important fibre source.

Keywords: acorn flour; gluten-free dough; fibre-rich ingredient; underexploited resources; rheology; pasting properties

---

1. Introduction

Gluten is one of the most important ingredients in bread making. It is responsible for the viscoelastic behaviour of the dough [1]. Gluten matrix has a major role on the main dough properties: extensibility, stretching resistance, mixing tolerance, gas-holding capacity and allows to obtain a high-quality bread crumb structure [2].

On the other hand, gluten-free dough (GFD) is unable to form a protein network similar to gluten. This raises several difficulties and thus the loss of quality on gluten-free bread baking [3]. The lack of gluten has a critical effect on dough rheology (less cohesive and elastic than wheat dough), formulation process and sensory quality of the final product [2]. The replacement of gluten in baking is a great challenge. There is no raw material or any other ingredient capable of completely replacing gluten, in terms of structural builder. Only the combination of different ingredients and its interactions, with adequate technologies, can improve gluten-free bread (GFB) quality [3–6]. Therefore, innovative ingredients with structural properties, novel approaches and processes, have been studied and tested with notable results so far obtained [7–11]. In our study, a mixture of natural gluten-free flours was used: buckwheat flour, rice flour, potato starch, acorn flour, and hydroxipropylmethylcellulose (HPMC) as thickening agent.

Rheology measurements are essential tools to understand the performance of the doughs, allowing to predict the final result and the before quality of bread [12]. Among the different rheological techniques, the small amplitude oscillatory shear (SAOS), demonstrated to be adequate to characterise viscoelastic materials and its structure properties [13]. The rheology of GFD, is also very important in order to understand the doughs behaviour during the baking process. Although the studies about this subject have recently increased, the information available is still limited [14]. Additionally, the majority of the GF flours are starch based, yet with different botanical sources, leading to completely diverse performances, according to the GF dough formulation [15].

Nowadays, due to resource scarcity worldwide, is essential to use by-products and underexploited natural raw materials that are not valued enough in food production. This fact led to market trends of sustainability [16], natural products and fibre-rich ingredients, which have contributed to the introduction of underexploited natural raw materials into food chain [17]. Acorn flour is an example, in particular to special requirements products as GF bread [18], increasing the importance of researching this flour. Several authors have been recently studying the incorporation of acorn flour in gluten containing biscuits and bread, in combination with wheat and barley. In all studies, the incorporation of acorn flour until a specific proportion improved the quality of the different final products, both in nutritional and sensorial characteristics [19–21]. Korus et al. [22] and Skendy et al. [11] have studied the effect of acorn flour addition in GF bread making, and both concluded that acorn flour incorporation in GF bread could be useful for nutritional and technological reasons. Additionally, many other innovative GF flours, and their impact on GF bread making, have been studied. Particularly, alternative sources of protein, like insects and legumes, found to improve GF bread nutritional composition and contribute to substitute gluten role [23,24]. In addition, microalgae, sorghum, millet, pseudo-cereals, fruit-based flours, by-products and other naturally GF flours have been tested, with very promising results [5,8,9].

Acorn is the fruit of *Quercus* genus tree. These trees are the basis of the Sustainable Agriculture System named "*Montado*" in South of Europe. In Portugal, acorns are abundant and can be considered an underexploited resource, since nearly 50% of the fruits in the trees are not used or harvested [17]. Many authors [25,26] state the use of acorn flour to bake bread has occurred since before the Roman Era. Veiga de Oliveira et al. [27] go further, referring that this flour was already consumed even before the Romans [27]. This fruit was traditionally present in the Mediterranean diet, some decades ago, especially in scarcity years [26]. The acorns most commonly used in human feed were from *Quercus ilex* and *Quercus rotundifolia* (holm oak). Since they are not so bitter comparing to other species, is not necessary to remove bitterness, as Korus et al. [22] described in their work. According to Silva and co-workers [18], acorn flour has interesting nutritional characteristics, and is rich in lipids (8.5%), especially in unsaturated fatty acids and fibre (9.5%–11.5%). Fibres have a major role in GFD rheology characteristics, besides improving nutritional GF bread quality [28,29].

Comparing this work with the referred authors [11,22] who have studied acorn flour incorporation in GF bread, it is important to highlight the innovation of our study. Since holm oak acorn is naturally slightly sweet, no bitterness removal treatment is needed, as mentioned before, which is a great technological advantage. The amount of water added to each flour blend was assessed through Micro-DoughLAB tests. Bread control formulation has buckwheat flour besides starch-based flours. Finally, pasting properties and dough TPA measurements were performed.

The objective of this work was to study the impact of acorn flour from holm oak, on the rheology characteristics of the GF dough, and also on the dietary fibre content, both soluble and insoluble fractions, of the GF baked breads. Two levels of incorporation (23% and 35%) were considered.

2. Materials and Methods

2.1. Raw Materials

Doughs formulations were prepared with the following flours: buckwheat flour (Próvida, Pêro Pinheiro, Portugal), acorn flour from dried holm oak (*Quercus ilex* and *Quercus rotundifolia*) (Terrius, Marvão, Portugal), rice flour (Fábricas Lusitana, Castelo Branco, Portugal), potato starch (Colmeia do Minho, Paio Pires, Portugal), dried yeast (Fermipan®, Lesaffre, Marcq-en-Baroeul, France), hydroxipropyl methylcellulose (HPMC) as a gelling agent (Wellence™ 321, Dow, MI, USA), sugar, sunflower oil, salt, and water. The flour ingredients and the HPMC were kindly supplied for free, except yeast, sugar, oil and salt that were purchased from local market.

2.2. Dough Preparation and Sampling

Fernandes [30] developed and analysed a GFD recipe that showed promising results, based on buckwheat flour, rice flour and potato starch, with HPMC as a gelling agent. The control dough (C) was prepared without acorn flour addition. Preliminary assays were conducted to adjust the water content of the control formulation to produce breads having the best quality, based on bread volume and crumb firmness, by testing different water hydrations (data not presented). Two incorporation levels of acorn flour were tested, 23% *w/w* (sample A23%) and 35% *w/w* (sample A35%) of total flours mixture, respectively, 50% and 75% in relation to the buckwheat flour. The three formulations studied are summarised in Table 1.

Table 1. Formulation of the gluten-free dough (GFD) samples and respective codes.

Ingredients (%)	Control (C)	Acorn 23% (A23%)	Acorn 35% (A35%)
Buckwheat flour	46.0	23.0	12.0
Rice flour	31.0	31.0	31.0
Potato starch	23.0	23.0	23.0
Acorn flour	0.0	23.0	35.0
Sunflower oil (in relation to flours)	5.5	5.5	5.5
HPMC (in relation to flours)	4.6	4.6	4.6
Dried yeast (in relation to flours)	2.8	2.8	2.8
Sugar (in relation to flours)	2.8	2.8	2.8
Salt (in relation to flours)	1.8	1.8	1.8
Water absorption (14% moisture basis)	65.0	63.0	62.0

Micro-doughLAB 2800 (Perten Instruments, Sidney, Australia) manufacturer's protocol at 63 rpm speed (AACC (American Association of Cereal Chemists) Method 54-21.01) was used to assess the optimum water absorption capacity for formulations with acorn flour. The peak value of torque of the optimised control-formulation was used as a reference (93 mN.m). Optimum water absorption values (amount of water needed to achieve target peak torque, corrected to 14% moisture basis) were determined for the formulations containing acorn flour in order to achieve a peak of 93 mN.m ± 4%.

The moisture of the different flours was measured through an automatic moisture analyser PMB 202 (Adam Equipment, Oxford, CT, USA). The values were buckwheat flour—12.35%; acorn flour—8.03%; rice flour—11.08%; potato starch—18.10%.

Ingredients were mixed in a Thermo processor equipment (Bimby—Vorwerk, Wuppertal, Germany), initially to activate the yeast, by adding water, yeast and sugar for 2 min at 27 °C, at velocity 1. Following, the other ingredients were added and mixed for 10 min in a dough mixing program (wheat ear symbol). The resulting dough was divided in three portions and poured into three equal containers and placed in a fermentation chamber Arianna XLT133 (Unox, Cadoneghe, Italy) for 50 min at 37 °C. All measurements were carried out after fermentation: three loaves of each formulation were prepared, and all the analyses were performed minimum in triplicate: three measurements were taken from independent three portions of each loaf. Mean values and standard deviations from each determination were recorded.

In a parallel work, the impact of acorn flour addition in the bread characteristics was studied, using exactly the same dough formulations above described. After fermentation, the dough was baked in an electric oven Johnson A60 (Johnson & Johnson, NJ, USA) for 50 min at 180 °C. Correlations between dough properties and breadmaking characteristics (bread firmness) were analysed.

2.3. Rheology Measurements

2.3.1. Pasting Properties

The pasting properties of flour blends, according to each GFD formulation, were tested using a Micro-doughLAB 2800 (Perten Instruments, Sidney, Australia), following the manufacturer's cooking protocol with some modifications [31]. Constant mixing rate at 63 rpm for 43 min (2580 s), and the following temperature cycle: 30 °C for 6 min, then rise temperature until 90 °C during 15 min, the temperature of 90 °C keeps constant for 7 min, then decrease to 50 °C during 10 min and finally, constant temperature of 50 °C for more 5 min, until the end. Only the flours and HPMC were added in the mixture, and 4.00 ± 0.01 g of each flour blend sample was weighed and placed in the chamber of Micro-doughLAB. The amount of water used was studied in previous mixing tests to find the optimum water absorption. Additionally, pure buckwheat flour and acorn flour were tested in the same conditions. Tests were performed at least in triplicate. Concerning rice flour and potato starch, since these ingredients were in the same share in all the formulations, were not submitted to pasting test.

2.3.2. Small Amplitude Oscillatory Measurements (SAOS) Testing

For dough characterisation, by dynamic rheometric measurements, formulations were prepared with all ingredients, as previously described. Rheological tests were performed using a controlled stress rheometer (MARS III, Haake, Germany) coupled with a UTC-Peltier system. A serrated parallel-plate P20 (diameter of 20 mm) was used, with a gap between plates of 1 mm. Each sample was prepared in a small portion of dough, and after 50 min of fermentation at 37 °C in a fermentation chamber Arianna XLT133 (Unox, Cadoneghe, Italy), was allowed to rest for 30 min at 5 °C. The fermented dough samples were placed between the plates, and after its adjustment, the edges were covered with paraffin to protect the sample from dehydration during measurements. Rheometer temperature was kept at 5 °C during test performance to avoid dough fermentation. Samples were left for 15 min before the test, to allow residual stress to relax and to stabilise the temperature. All rheological measurements were run, at least, in triplicate.

Small amplitude oscillatory shear (SAOS) tests were conducted to assess linear viscoelastic properties of the GFD. The dynamic linear viscoelastic region of each sample was previously determined through stress sweep tests at frequency of 6.28 rad/s (1 Hz). Frequency sweeps (from 0.0628 to 628 rad/s) were performed at constant shear stress of 10 Pa.

2.3.3. Dough Texture Profile Analysis (TPA)

Texture profile analysis (TPA) was performed after dough fermentation at 37 °C during 50 min, using a texturometer TA.XT.plus (Stable Micro Systems, Surrey, England) equipped with a load cell of 5 kg and penetration mode of the equipment set up. Dough samples were placed in cylindric containers with 100 mm in diameter and a height of 80 mm. These cylinders were completely filled by the sample.

The measurement conditions were 19 mm diameter acrylic cylindrical probe, 20 mm of penetration distance, 1 mm/s of crosshead speed and 5 s of waiting time between the two measurements cycles. Each measurement was repeated at least four times. Firmness and cohesiveness were the main representative texture parameters obtained from TPA, to characterise the dough, as it was reported by Nunes et al. [8].

2.4. Colour Characterisation and pH Values of GF Dough

Dough colour was determined using a colorimeter Croma-Meter CR 400 (Konica-Minolta Sensing Americas, New Jersey, USA), through the CIE L*a*b* system (International Commission on Illumination) using the following parameters: L*—lightness variable (L* = 100 white, L* = 0 black); a*—intensity of green (−60 < a* < 0) or red (0 < a* < +60); and b*—intensity of blue (−60 < b* < 0) or yellow (0 < b* < +60). Each sample was measured in triplicate.

pH measurements were performed in three different points of each fermentation dough, with a potentiometer pH-Meter Basic 20 (Crison Instruments, Barcelona, Spain), until stabilisation.

2.5. Dietary Fibre: Soluble, Insoluble and Total Fibre in GF Bread

According to Martins et al. [32], gluten-free breads were prepared from the same GF doughs studied in this work, as previously described. The soluble, insoluble and total fibre of GF breads (Control, A23% and A35%), were determined by using a "Total Dietary Fibre Assay Kit" from Megazyme (Wicklow, Ireland). According to manufacturer's information, the analysis can be considered to be divided in two methods: (1) based on AOAC (Association of Official Agricultural Chemists) Method 991.43 "Total, Soluble, and Insoluble Dietary Fibre in Foods" (First Action 1991) and AACC Method 32-07.01 "Determination of Soluble, Insoluble, and Total Dietary Fibre in Foods and Food Products" (Final Approval 10-16-91); (2) determination of total dietary fibre based on AACC method 32-05.01 and AOAC Method 985.29. All procedures were done at least in duplicate.

2.6. Statistical Analysis

Results were analysed using the statistical programme Origin Pro 8 (OriginLab Corporation, MA, USA). Experimental data were compared using analysis of variance (one-way ANOVA), and the Tukey test was used to evaluate mean differences at a confidence level of 95%, with significant differences considered to be $p < 0.05$. Results are presented as mean values and standard deviations. Correlations were tested between the different variables, using the program Statistica 10.0 and the function Bivariate Scatterplot.

3. Results and Discussion

3.1. Dough Rheology Characterisation

3.1.1. Pasting Properties

Flour blends' and pure flours' pasting curves, obtained from the MidrodoughLab are represented in Figure 1, with the main points marked with a blue circle.

In Figure 1, it is possible to observe the maximum torque of the flour blends' mixing curves, that is represented in C1, with a constant temperature of 30 °C. After 6 min, the temperature starts rising. The graphic shows the breakdown (C2), minimum torque (linked with minimum viscosity) at 61 °C, corresponding to protein weakening. All the flour blends had a similar response concerning both temperature and torque (40 m.Nm). At the same stage (C2), pure flours showed the minimum torque, related with a minimum of viscosity, at the same temperature, but with a lower torque comparing with the blends. Buckwheat flour torque (directed related with its viscosity) was around 20 m.Nm and acorn flour dropped to 1 m.Nm, which is in agreement with [33], that found likewise, lowest viscosity in the breakdown for acorn starch comparing with buckwheat.

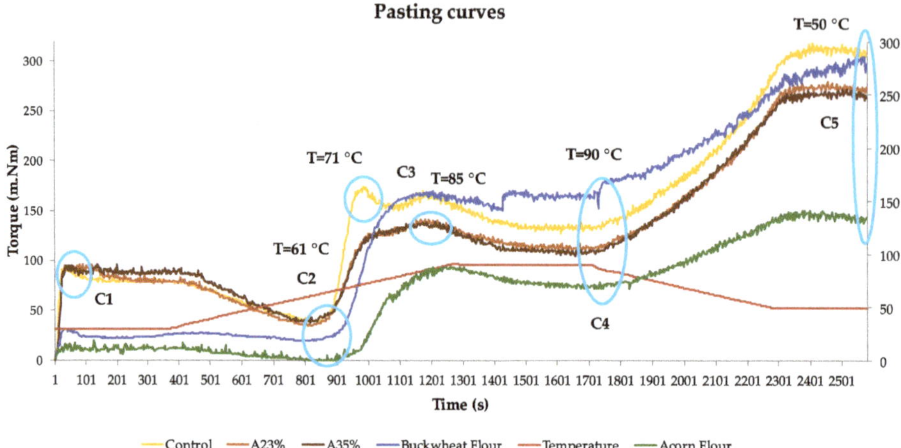

Figure 1. Pasting curves of the tested blend flours: control, A23% and A35%, and also buckwheat and acorn flour.

Along the heating process, the starch granules retain water and swell, which results in viscosity increasing, until peak viscosity (C3) is reached and indicates water-binding capacity, and starch gelatinisation [34,35]. Control flours mixture reached the gelatinisation peak torque (175 m.Nm) at 71 °C and both acorn flour mixtures at 85 °C, but with a lower torque (137 m.Nm). From the curve of acorn flour, with a lower peak torque (87 m.Nm), it is possible to understand this behaviour, where the contribution of acorn flour decreased the viscosity of the blends A23% and A35%. Buckwheat flour presents higher viscosity, as well as the control blend, where this flour is present in 46% share. In GFD, starch gelatinisation performs an essential function, since it has the aptitude to trap gas bubbles in its own matrix, promoting the gas holding capacity of the dough [36,37].

Starch gelatinisation depends on the starch characteristics and its swelling capacity. Moreover, the starches establish interactions with the HPMC, a fact that also contributes to the maximum viscosity of the dough, reached during gelatinisation [15]. According to Hager et al. [38], buckwheat flour has approximately 60% of starch, with around 25% amylose, while acorn flour has around 50% of starch [18], where 55% is amylose [39]. Singh et al. [40] state that amylose and lipid contents are one of the causes for the differences between the swelling capacity of starches. Higher lipid content is supposed to decrease the swelling capacity of individual granules [40]. However, Debet and Gindley [41] suggest that when the amylose content is high, the effect of lipids is secondary, concluding that high amylose content inhibits swelling. Correia et al. [39], concluded fewer amylose leaching and lower degree of gelatinisation has been attributed to acorn structure type of starch. Regarding the higher temperature of acorn flour gelatinisation, according to Correia et al. [39], indicates increased resistance to swelling, a characteristic response of acorn starch. On the contrary, according to Debet and Gindley [41], higher viscosity reveals starch with higher swelling capacity, and Torbica et al. [42] identify it as a pasting property of buckwheat flour.

The following stage is characterised by a viscosity decreasing to a minimum (breakdown). This happens when the granule absorbs as much water as to achieve its rupture point, while temperature keeps increasing [40]. As we can see from the results, the breakdown (C4) occurs at the end of 90 °C with higher torque (132 m.Nm) for control dough and lower torque (110 m.Nm), meaning lower viscosity, for both acorn blends. Buckwheat flour showed the highest torque (154 m.Nm) and acorn the lowest (68 m.Nm), revealing the contribution that each pure flour has in the respective blends, according to the different levels of incorporation in the dough.

When gelatinised starch temperature cools, viscosity increases until the end of the test (C5), the moment when we can see the formation of a gel, resulting from amylose retrogradation [34]. Regarding Figure 1, is possible to observe that retrogradation torque was quite similar between control blend (305 m.Nm) and pure buckwheat flour (300 m.Nm). Following, the acorn flour blends, present a lower torque, respectively, A23% (270 m.Nm) and A35% (262 m.Nm). Finally, pure acorn flour torque was the lowest (129 m.Nm). Despite this retrogradation low viscosity of acorn flour, it is possible to recognise that acorn flour blends have a quite high viscosity, when comparing with the pure flour. Synergetic effects between the starches, proteins, HPMC and fibres, and also the proportion of acorn flour in the blends, can explain why these blends are able to maintain a higher viscosity in the retrogradation process. Paste retrogradation is influenced by the amylose content and also by the structural arrangement of the starch chains [37,38]. According to Hager et al. [35], buckwheat starch presents a granular shape, whereas acorn starch granules show a spheroid/ovoid and cylindrical shape [43].

Since our study is focused on acorn flour, it is important to highlight that among the different species of oak trees, the acorn starch is also different due to taxonomic characteristics of the plant [43]. For this reason, to make possible the comparison about starch behaviour, we used the studies based on the same species: *Quercus rotundifolia* and *Quercus ilex* from Cappai et al. [43] and Correia et al. [39].

Finally, it is also important to mention the role that proteins have in the dough rheology behaviour, contributing as well to gluten replacing functions [3]. Different authors [41,44,45] state that interactions between the different types of starch and proteins, have an influence on the dough rheology characteristics, affecting also pasting properties of starch. Concerning the developed doughs, the average protein content (data obtained from suppliers) of buckwheat is 12%, and acorn flour is 4.5%, in accordance with literature [18,38]. As previously described in Section 2.2, samples A23% and A35% have 50% and 75% replacement of buckwheat flour, by acorn flour, respectively, when comparing with the control sample. This means that buckwheat flour changes from 46% incorporation in control to 23% in A23% sample and to 11% in A35%, with a corresponding increasing of 23% and 35% of acorn flour. In relation to protein, acorn flour presents approximately 8% less content than buckwheat flour, resulting in a lower protein content on sample blends A23% and A35%. This will influence protein–starch interactions and consequently pasting properties. As we can observe from the graphic, torque values in C3, C4, and C5, are lower in acorn flour blends when comparing to the control blend. Similar results have been reported in different studies about proteins influence in starch pasting properties [46–48].

3.1.2. Small Amplitude Oscillatory Measurements (SAOS)

The mechanical spectra of the GF doughs (control, A23%, A35%) are presented in Figure 2A. The mechanical spectra of GF doughs represent the variation of storage modulus (G′) and loss modulus (G′′) as a function of angular frequency (ω). In order to promote a more systematic comparison of the spectra, the values of G′ obtained at 62.8 rad/s (10 Hz) are presented in Figure 2B.

It is possible to observe in Figure 2A, that both moduli G′ and G′′ show high frequency dependence. The three GF doughs compared presented higher values of G′ in comparison with G′′, in the whole frequency ranges, expressing a predominance of the elastic behaviour. It is also important to note that the control dough shows lower values of the viscoelastic functions and a pseudo-terminal region at low frequency values, with a tendency for a crossing point at very low frequencies. This type of behaviour reflects the existence of a less structured system in the control dough, which may be associated with a less stable structure [22]. In other words, the incorporation of acorn flour had an important role in increasing the degree of structuring of the dough. This increase in the degree of structure should have resulted from the greater ability of acorn starch to create structures, which was evidenced by the results obtained in terms of pasting curves. Our results are in accordance with different authors [22,49,50] that also have studied GF dough with different botanical source flours (respectively, acorn flour, carob flour and legume flours), and obtained similar results, where G′ > G′′, showing a soft gel-like structure.

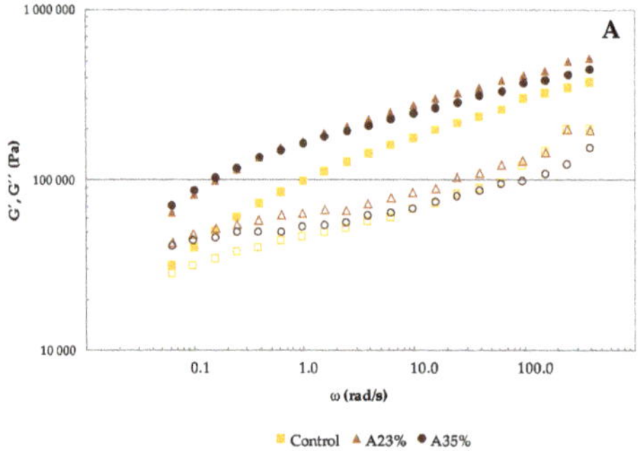
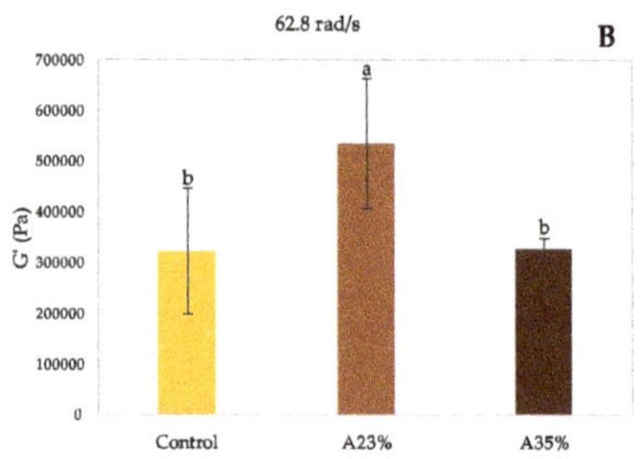

Figure 2. (A): Mechanical spectra of GF doughs (control, A23%, A35%); G′ (storage modulus—filled symbol), G′′ (loss modulus—open symbol). **(B)**: G′ values at 62.8 rad/s (10 Hz) for the same samples.

About G′ values at 62.8 rad/s, a significantly higher value ($p < 0.05$) for A23% dough was observed, meaning that the addition of acorn flour in dough formulation induced a significantly increase of elastic modulus, acting as a strengthening agent of the structure. However, the dough with 35% of acorn flour, reduces significantly G′ values, being similar to control dough. A similar effect will be found in the texture results, in terms of the TPA parameters. This fact indicates that the growing incorporation level of acorn flour presents a negative impact on the structure, since it is possible to observe a modification of the rheological behaviour, with the reduction of the elastic modulus at 62.8 rad/s. The aforementioned behaviour is not significantly different ($p > 0.05$) from the one of control dough, showing that the benefits, in terms of viscoelastic characteristics, induced from the addition of acorn flour are not extend to higher levels of incorporation. Regarding the improvement of rheology characteristics observed in acorn flour doughs, it is also important to underline the contribution of the fibre content present in acorn flour. Several authors have studied the impact of fibre in GFD formulation, with the objective of improving rheology. The dough formulation with the inclusion of fibre-rich ingredients, positively influenced rheological properties, with higher values of both moduli, and structuring the GF dough closer to a gel-like material [11,28,35,51,52].

Korus et al. [22] obtained different results, where partial (20%, 40%, 60%) replacement of starch with acorn flour, resulted in a significant increase in both moduli, and proportional to the extent of acorn flour replacement. Nevertheless, it is important to clarify that the control dough formulation was based on corn and potato starch, totally different from the control dough in our study. Nevertheless, the author refers that higher levels of acorn flour incorporation revealed a negative impact, since it caused the reduction of loaf volume, not shown in rheology tests [22]. This should also be related with a negative impact on the dough viscoelasticity.

3.1.3. Texture Profile Analysis (TPA)

Figure 3 shows the results of TPA: firmness and cohesiveness. The incorporation of both acorn flour levels, 23% and 35%, increased significantly ($p < 0.05$) the firmness of the dough comparing with control. Between the two acorn flour doughs, the firmness is significantly ($p < 0.05$) higher in A23% when comparing with A35%. For the lowest level of incorporation, it was possible to obtain a more structured dough, already supported in terms of increasing the elastic modulus (G′), which corresponds to a higher firmness value. For the 35% of acorn flour incorporation, the extension of the network developed is weaker, resulting in a softer dough, similar to the control. Probably, the excess of acorn starch caused the weakening of this network, as we can observe from pasting curve analysis, where a noticeable decrease of acorn flour viscosity parameters is revealed.

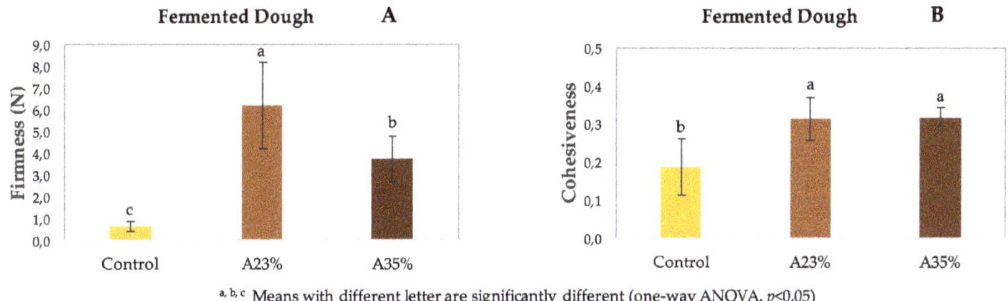

Figure 3. Firmness (**A**) and cohesiveness (**B**) of GF doughs (control, A23% and A35%).

These findings are in accordance with the previous rheology measurements results discussion. Nevertheless, the firmness of the dough with 35% of acorn flour is significantly higher than the control dough. The results show that partial replacement of buckwheat flour by acorn flour improves the firmness of the dough, for either levels of acorn flour.

Cohesiveness is associated with the level of structure between the different elements that take part in the dough matrix [53]. The obtained results demonstrate that dough A23% and also dough A35% significantly increased ($p < 0.05$) their cohesiveness, when compared with control dough. However, no significant differences were found between the two levels of acorn flour incorporation. From these results, it is possible to observe that the incorporation of acorn flour in the formulation, with the reduction of buckwheat flour share, increases the cohesiveness of the dough. Furthermore, Correia et al. [39] also concluded that acorn starches have the ability to contribute to a coherent structure.

The increase of firmness and cohesiveness of the dough with acorn flour addition, together with the reduction of buckwheat flour incorporation level, can be explained due to the nutritional composition of acorn flour [18] in comparison to buckwheat flour [38]. As previously mentioned, acorn flour has a higher content of fibre (9.5–11.5%) and total fat (8.5%), when compared to buckwheat flour with about 2% of fibre and 5% of total fat [18,38]. According to different authors presented in a review article, the level of fibre has a major role as structure enhancer in GF leavened bread, as well as the interactions between the different components of the complex dough system. Due to water binding capacity, gel forming ability, fat mimetic, textural and thickening effects, fibres have a positive impact on the texture, improving both firmness and cohesiveness [52]. As previously reported by Graça et al. [7], firmness and cohesiveness are the most representative texture parameters of GF dough, as they have a great influence on bread quality and acceptance [3].

3.2. Impact of Acorn Flour Addition in Bread Texture

In a parallel work, the impact of acorn flour addition in the bread texture was characterised, using exactly the same formulations and processing conditions used in the present study. In that work, it was found that acorn flour incorporation, generally, induces an increase in bread firmness. Through the results obtained in the present study, it was possible to establish a positive correlation between the bread firmness and the viscoelastic characteristics of the dough—G′ at 0.628 rad/s (Figure 4A) and between the bread firmness and the dough firmness (Figure 4B).

The obtained correlations, which allow to predict the behaviour of the bread, in terms of firmness, taking into account the dough rheological behaviour and its texture, are extremely useful in terms of the gluten-free bakery industry. Similar correlations have been obtained by other authors [7,37].

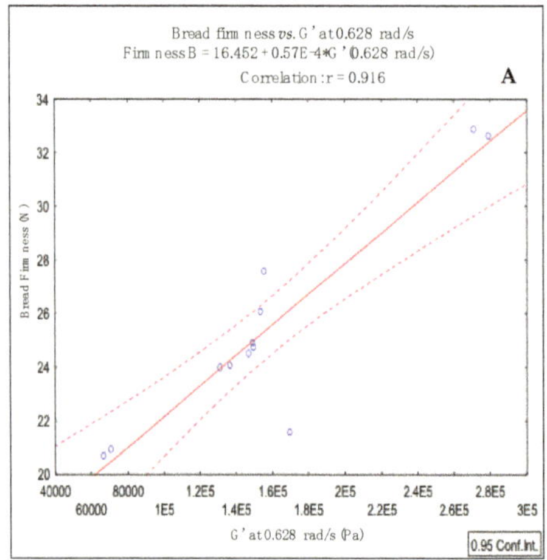

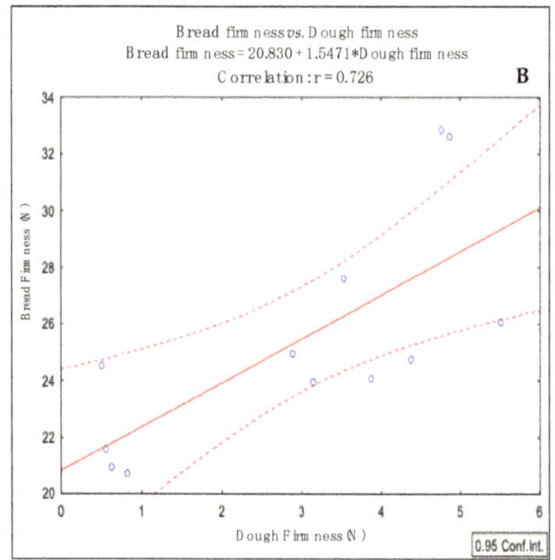

Figure 4. (**A**): Correlation between bread firmness and the viscoelastic characteristics of the dough (G′ at 0.628 rad/s) and (**B**): correlation between the bread firmness and dough firmness.

3.3. Characteristics of GF Dough: Colour and pH

Table 2 presents the results of GF dough colour measurements and respective pH. It can be observed that pH significantly decreased from control dough to acorn dough. Either levels of added acorn flour led to an acidification of the dough, although there are no significant differences between A23% and A35% ($p > 0.05$).

Table 2. Colour parameters and pH of GF doughs (control, A23% and A35%).

	Control	A23%	A35%
pH	5.35 [a] ± 0.04	5.09 [b] ± 0.05	5.05 [b] ± 0.05
Colour L*	82.49 [a] ± 0.42	70.90 [b] ± 0.08	66.42 [c] ± 0.26
Colour a*	0.56 [c] ± 0.21	6.11 [b] ± 0.35	7.89 [a] ± 0.31
Colour b*	15.50 [c] ± 0.33	24.45 [b] ± 0.83	27.31 [a] ± 0.65

Mean values with different letters in the same row are significantly different (one-way ANOVA, $p < 0.05$).

Concerning dough colour, it is possible to observe that L* value has reduced significantly ($p < 0.05$) with the increase of acorn flour content. These results show a reduction in whiteness and an increase in the browning index, meaning the darkness of the dough. This effect on the dough is important, as it will contribute to the improvement of bread colour. This is a relevant parameter when observing consumers' preferences. For GF bread, this characteristic is even more important, since it frequently presents pale colours when compared to their wheat counterparts, what is referred as a depreciative factor in GF bread [2]. Regarding the a* and b* parameters of the dough, it is possible to observe a significant ($p < 0.05$) increase with acorn flour incorporation, which means, respectively, a more intensive red than green, and a predominance of yellow over blue. Colour differences (ΔE) between control bread and acorn breads can be calculated by the following equation:

$$\Delta E^* = (\Delta L^{*2} + \Delta a^{*2} + \Delta b^{*2})^{1/2} \tag{1}$$

The human eye can only detect the difference in the colours if ΔE > 5 [54]. When comparing the control dough with both acorn flour dough, the obtained results for were ΔE*(control–A35%) = 21.25; and ΔE*(control–A23%) = 15.66, revealing that the difference between the dough colours was really high. Finally, the difference between the two levels of acorn flour dough was ΔE^*(A23%–A35%) = 5.61, almost in the human eye threshold perception.

3.4. Dietary Fibre: Soluble, Insoluble and Total Fibre

As fibre content increases, besides the rheology improvement above mentioned, is also possible to improve GF bread nutritional profile. Several authors have studied the addition of fibre-rich raw materials in GF bread formulation, obtaining very promising results [52]. Fratelli et al. [55] evaluated and optimised the applications of fibre (psyllium) in GF formulation, and obtained a final recipe with an increase of fibre content from 2.5% to 4% and a decrease of glycaemic response, with good acceptability scores. Rocha Parra et al. [35,51] studied apple pomace as an alternative source for fibre enrichment of GF baked products. Tsatsaragkou et al. [49] concluded that carob flour could be a promising solution to develop high quality formulations of GF bread with a fibre-rich ingredient. Hager et al. [38] analysed the chemical profile of nutrient-dense GF flours, and compared them with wheat flours. In addition, Hager et al. [38] have also studied soluble and total fibre, together with recommended daily intake.

According to AACC [56], dietary fibre (DF) is defined as the carbohydrates (from edible part of plants) which are resistant to digestion and adsorption in the human small intestine, with complete or part fermentation in the large intestine. The term DF comprises polysaccharides, oligosaccharides and associated plant compounds. DF is fractionated into insoluble fibre, linked with intestinal regulation, and soluble fibre associated to serum cholesterol reduction levels and with glycaemic response [38,56].

According to Hager et al. [38], buckwheat flour presents approximately 2% of DF, from which about 23% is soluble fibre. Concerning acorn flour, Silva et al. [18] obtained values between 9.5% and 11.5%. Our results were also similar, with 10% of DF, from which about 9% is soluble fibre (data not shown).

The results of GF bread fibre analysis, soluble fraction, insoluble fraction and total fibre are presented in Table 3. From the presented data, it is possible to observe that the incorporation of both levels of acorn flour increased significantly ($p < 0.05$) the total fibre and insoluble fibre, comparing with control bread formulation. This was expected, since acorn is a good source of DF, particularly in the insoluble fraction [22].

Table 3. Insoluble, soluble and total fibre of GF breads (in dry matter): control, A23% and A35%.

	Control	**A23%**	**A35%**
Insoluble Fibre (%)	8.99 [b] ± 0.63	11.96 [a] ± 0.53	12.46 [a] ± 0.92
Soluble Fibre (%)	1.32 [a] ± 0.38	0.31 [b] ± 0.24	0.12 [b] ± 0.10
Total Fibre (%)	10.31 [b] ± 0.65	12.27 [a] ± 0.75	12.58 [a] ± 0.90

Mean values with different letters in the same row are significantly different (one-way ANOVA, $p < 0.05$).

However, it is important to note that the soluble fibre content of the breads significantly decreased ($p < 0.05$), comparing with the control formulation, when acorn flour incorporation was considered. This fact can be explained since the soluble fibre fraction of acorn flour in comparison to buckwheat flour is much lower.

According to Melini and Melini [57], a healthy diet needs to be diverse, balanced and it must assure a high intake of DF. Even though DF health benefits are profusely documented, is frequently concluded that recommended daily intake is not followed by the majority of the consumers [35]. In the case of consumers with special requirements as GF diet, it is even more difficult, since GF products are in general described with a lack of fibre [58].

Following USDA recommendations, the DF Reference Intake is 25 g/day for women and 38 g/day for men. If we consider an ingestion of 100 g of fresh bread per day, the 23% acorn flour bread would serve 7.5 g of fibre. This represents about 30% and 20% of daily adequate DF intake for women and men, respectively, meaning that acorn flour bread can be an important vehicle of fibre, with the possibility to use the respective nutritional claim.

Thus, acorn flour seems to be a very promising ingredient, in order to improve DF content in GF bread.

4. Conclusions

Based on the obtained results, it can be stated that acorn flour significantly affects the rheology properties of the doughs. Moreover, acorn flour has an impact on the dough's mixing and pasting curves and improved the firmness, the cohesiveness and the viscoelasticity of the fermented dough. The 23% of acorn flour incorporation presented better results when comparing with 35%. According to SAOS, GF doughs exhibit a weak gel-elastic like behaviour with G' values higher than G" and frequency dependent. Acorn flour incorporation caused the acidification and increased the darkness of the dough, that will have a positive impact in terms of sensory appreciation of the bread. Thus, acorn flour can be a very promising ingredient, in order to improve both rheological GF dough properties and nutritional GF bread quality, in particular DF content, a really important nutrient in special requirement diets.

Author Contributions: R.B.M. and L.M.M.F. conceived and planned the fibre analysis experiments. R.B.M., M.C.N. and A.R. conceived and planned all the other experiments. R.B.M. developed all the sample preparations and laboratory analyses; performed the data analysis, interpretation of the results, and wrote the paper draft. M.C.N., A.R. and L.M.M.F., supervised the research work. M.C.N., A.R., L.M.M.F., J.A.P. and A.I.R.N.A.B. contributed to the discussion of the data and revised the manuscript. All authors have read and agreed to the published version of the manuscript.

Funding: This work had the financial support provided by the FCT-Portuguese Foundation for Science and Technology PD/BD/135332/2017 under the Doctoral Programme "Agricultural Production Chains—from fork to farm", through the research unit UIDB/04033/2020 (CITAB), UID/AGR/04129/2013 (LEAF) and UIDB/00616/2020 (CQVR).

Acknowledgments: The authors would like to acknowledge the companies Próvida, Fábricas Lusitania, A Colmeia do Minho, and Dow that kindly provided the raw materials used in this study.

Conflicts of Interest: The authors declare no conflict of interest.

References

1. Schofield, J.D. Wheat proteins: Structure and functionality in milling and breadmaking. In *Wheat Production, Properties and Quality*, 1st ed.; Bushuk, W., Rasper, V.F., Eds.; Springer: Boston, MA, USA, 1994; pp. 73–106. [CrossRef]
2. Gallagher, E.; Gormley, T.R.; Arendt, E.K. Recent advances in the formulation of gluten-free cereal-based products. *Trends Food Sci. Technol.* **2004**, *15*, 143–152. [CrossRef]
3. Matos, M.E.; Rosell, C.M. Understanding gluten-free dough for reaching breads with physical quality and nutritional balance. *J. Sci. Food Agric.* **2015**, *95*, 653–661. [CrossRef] [PubMed]
4. Rai, S.; Kaur, A.; Chopra, C.S. Gluten-Free Products for Celiac Susceptible People. *Front. Nutr.* **2018**, *5*, 116. [CrossRef] [PubMed]
5. Padalino, L.; Conte, A.; Del Nobile, M. Overview on the General Approaches to Improve Gluten-Free Pasta and Bread. *Foods* **2016**, *5*, 87. [CrossRef] [PubMed]
6. Deora, N.S.; Deswal, A.; Mishra, H.N. Alternative Approaches Towards Gluten-Free Dough Development: Recent Trends. *Food Eng. Rev.* **2014**, *6*, 89–104. [CrossRef]
7. Graça, C.; Raymundo, A.; Sousa, I. Yogurt as an Alternative Ingredient to Improve the Functional and Nutritional Properties of Gluten-Free Breads. *Foods* **2020**, *9*, 111. [CrossRef]
8. Nunes, M.C.; Graça, C.; Vlaisavljević, S.; Tenreiro, A.; Sousa, I.; Raymundo, A. Microalgal cell disruption: Effect on the bioactivity and rheology of wheat bread. *Algal Res.* **2020**, *45*, 101749. [CrossRef]
9. Torbica, A.; Belovic, M.; Tomic, J. Novel breads of non-wheat flours. *Food Chem.* **2019**, *282*, 134–140. [CrossRef]
10. Vidaurre-Ruiz, J.; Matheus-Diaz, S.; Salas-Valerio, F.; Barraza-Jauregui, G.; Schoenlechner, R.; Repo-Carrasco-Valencia, R. Influence of tara gum and xanthan gum on rheological and textural properties of starch-based gluten-free dough and bread. *Eur. Food Res. Technol.* **2019**, *245*, 1347–1355. [CrossRef]
11. Skendi, A.; Mouselemidou, P.; Papageorgiou, M.; Papastergiadis, E. Effect of acorn meal-water combinations on technological properties and fine structure of gluten-free bread. *Food Chem.* **2018**, *253*, 119–126. [CrossRef]

12. Stojceska, V.; Butler, F. Investigation of reported correlation coefficients between rheological properties of the wheat bread doughs and baking performance of the corresponding wheat flours. *Trends Food Sci. Technol.* **2012**, *24*, 13–18. [CrossRef]

13. Dobraszczyk, B.J.; Morgenstern, M.P. Rheology and the breadmaking process. *J. Cereal Sci.* **2003**, *38*, 229–245. [CrossRef]

14. Masure, H.G.; Fierens, E.; Delcour, J.A. Current and forward looking experimental approaches in gluten-free bread making research. *J. Cereal Sci.* **2016**, *67*, 92–111. [CrossRef]

15. Zhang, D.; Mu, T.; Sun, H. Comparative study of the effect of starches from five different sources on the rheological properties of gluten-free model doughs. *Carbohydr. Polym.* **2019**, *176*, 345–355. [CrossRef] [PubMed]

16. Notarnicola, B.; Tassielli, G.; Renzulli, P.A.; Monforti, F. Energy flows and greenhouses gases of EU (European Union) national breads using an LCA (Life Cycle Assessment) approach. *J. Clean. Prod.* **2017**, *140*, 455–469. [CrossRef]

17. Vinha, A.F.; Costa, A.S.G.; Barreira, J.C.M.; Pacheco, R.; Oliveira, M.B.P.P. Chemical and antioxidant profiles of acorn tissues from *Quercus* spp.: Potential as new industrial raw materials. *Ind. Crops Prod.* **2016**, *94*, 143–151. [CrossRef]

18. Silva, S.; Costa, E.M.; Borges, A.; Carvalho, A.P.; Monteiro, M.J.; Pintado, M.M.E. Nutritional characterization of acorn flour (a traditional component of the Mediterranean gastronomical folklore). *Food Meas.* **2016**, *10*, 584–588. [CrossRef]

19. Švec, I.; Hrušková, M.; Kadlčíková, I. Features of flour composites based on the wheat or wheat-barley flour combined with acorn and chestnut. *Croat. J. Food Sci. Technol.* **2018**, *10*, 89–97. [CrossRef]

20. Pasqualone, A.; Makhlouf, F.Z.; Barkat, M.; Difonzo, G.; Summo, C.; Squeo, G.; Caponio, F. Effect of acorn flour on the physico-chemical and sensory properties of biscuits. *Heliyon* **2019**, *5*, e02242. [CrossRef]

21. Hruskova, M.; Svec, I.; Kadlcíková, I. Effect of chestnut and acorn flour on wheat/wheat-barley flour properties and bread quality. *Int. J. Food Stud.* **2019**, *8*, 41–57. [CrossRef]

22. Korus, J.; Witczak, M.; Ziobro, R.; Juszczak, L. The influence of acorn flour on rheological properties of gluten-free dough and physical characteristics of the bread. *Eur. Food Res. Technol.* **2015**, *240*, 1135–1143. [CrossRef]

23. Cappelli, A.; Oliva, N.; Bonaccorsi, G.; Lorini, C.; Cini, E. Assessment of the rheological properties and bread characteristics obtained by innovative protein sources (*Cicer arietinum*, *Acheta domesticus*, *Tenebrio molitor*): Novel food or potential improvers for wheat flour? *LWT* **2020**, *118*, 108867. [CrossRef]

24. Kowalczewski, P.Ł.; Walkowiak, K.; Masewicz, Ł.; Bartczak, O.; Lewandowicz, J.; Kubiak, P.; Baranowska, H.M. Gluten-Free Bread with Cricket Powder—Mechanical Properties and Molecular Water Dynamics in Dough and Ready Product. *Foods* **2019**, *8*, 240. [CrossRef] [PubMed]

25. Jacob, H.E. *6000 Anos de Pão*, 1st ed.; Antígona: Lisboa, Portugal, 2003; p. 587.

26. Ribeiro, O. *Portugal o Mediterrâneo e o Atlântico*, 4th ed.; Livraria Sá da Costa Editora: Lisboa, Portugal, 1986; p. 188.

27. Veiga de Oliveira, E.; Galhano, F.; Pereira, B. *Tecnologia Tradicional Portuguesa-Sistemas de Moagem*; INIC, Centro de Estudos de Etnologia: Lisboa, Portugal, 1983; p. 520.

28. Kiumarsi, M.; Shahbazi, M.; Yeganehzad, S.; Majchrzak, D.; Lieleg, O.; Winkeljann, B. Relation between structural, mechanical and sensory properties of gluten-free bread as affected by modified dietary fibers. *Food Chem.* **2019**, *277*, 664–673. [CrossRef] [PubMed]

29. Naqash, F.; Gani, A.; Gani, A.; Masoodi, F.A. Gluten-free baking: Combating the challenges-A review. *Trends Food Sci. Technol.* **2017**, *66*, 98–107. [CrossRef]

30. Fernandes, I.C.X. *Desenvolvimento de pães sem glúten enriquecidos com Tetraselmis chuii. Dissertação de Mestrado em Engenharia Alimentar*; Instituto Superior de Agronomia da Universidade de Lisboa: Lisboa, Portugal, 2019.

31. Dang, J.M.C.; Bason, M.L. Comparison of Old and New Dough Mixing Methods and their Utility in Predicting Bread Quality. In Proceedings of the 65th Australian Cereal Chemistry Conference, Coogee, NSW, Australia, 16–18 September 2015.

32. Martins, R.B.; Nunes, M.C.; Peres, J.A.; Barros, A.I.R.N.A.; Raymundo, A. Acorn Flour as bioactive compounds source in gluten free bread. In Proceedings of the Book of Abstracts of XX EuroFoodChem Congress, Porto, Portugal, 17–19 June 2019.

33. Cho, A.; Kim, S.K. Particle size distribution, pasting pattern and texture of gel of acorn, mungbean, and buckwheat starches. *Korean J. Food Sci. Technol.* **2000**, *32*, 1291–1297.

34. Sciarini, L.S.; Ribotta, P.; León, A.; Pérez, G. Influence of Gluten-free Flours and their Mixtures on Batter Properties and Bread Quality. *Food Bioprocess Technol.* **2010**, *3*, 577–585. [CrossRef]

35. Rocha Parra, A.F.; Ribotta, P.D.; Ferrero, C. Starch–Apple Pomace Mixtures: Pasting Properties and Microstructure. *Food Bioprocess Technol.* **2015**, *8*, 1854–1863. [CrossRef]

36. Abdel-Aal, E.-S.M. Functionality of Starches and Hydrocolloids in Gluten-Free Foods. In *Gluten-Free Food Science and Technology*; Gallagher, E., Ed.; Wiley-Blackwell: Oxford, UK, 2009; pp. 200–224.

37. Martínez, M.M.; Gómez, M. Rheological and microstructural evolution of the most common gluten-free flours and starches during bread fermentation and baking. *J. Food Eng.* **2017**, *197*, 78–86. [CrossRef]

38. Hager, A.S.; Wolter, A.; Jacob, F.; Zannini, E.; Arendt, E.K. Nutritional properties and ultra-structure of commercial gluten free flours from different botanical sources compared to wheat flours. *J. Cereal Sci.* **2012**, *56*, 239–247. [CrossRef]

39. Correia, P.R.; Nunes, M.C.; Beirão-da-Costa, M.L. The effect of starch isolation method on physical and functional properties of Portuguese nut starches. II. *Q. rotundifolia Lam. and Q. suber Lam. acorns starches*. *Food Hydrocoll.* **2013**, *30*, 448–455. [CrossRef]

40. Singh, N.; Singh, J.; Kaur, L.; Sodhi, N.S.; Gill, B.S. Morphological, thermal and rheological properties of starches from different botanical sources. *Food Chem.* **2003**, *81*, 219–231. [CrossRef]

41. Debet, M.R.; Gidley, M.J. Three classes of starch granule swelling: Influence of surface proteins and lipids. *Carbohydr. Polym.* **2016**, *64*, 452–465. [CrossRef]

42. Torbica, A.; Hadnađev, M.; Dapčević, T. Rheological, textural and sensory properties of gluten-free bread formulations based on rice and buckwheat flour. *Food Hydrocoll.* **2010**, *24*, 626–632. [CrossRef]

43. Cappai, M.G.; Alesso, G.A.; Nieddu, G.; Sanna, M.; Pinna, W. Electron microscopy and composition of raw acorn starch in relation to in vivo starch digestibility. *Food Funct.* **2013**, *4*, 917–922. [CrossRef]

44. Horstmann, S.W.; Lynch, K.M.; Arendt, E.K. Starch Characteristics Linked to Gluten-Free Products. *Foods* **2017**, *6*, 29. [CrossRef]

45. Schirmer, M.; Jekle, M.; Becker, T. Starch gelatinization and its complexity for analysis. *Starch-Stärke* **2015**, *67*, 30–41. [CrossRef]

46. Alonso-Miravalles, L.; O'Mahony, J.A. Composition, Protein Profile and Rheological Properties of Pseudocereal-Based Protein-Rich Ingredients. *Foods* **2018**, *7*, 73. [CrossRef]

47. Storck, C.R.; Zavareze, E.R.; Gularte, M.A.; Elias, M.C.; Rosell, C.M.; Dias, A.R.G. Protein enrichment and its effects on gluten-free bread characteristics. *LWT* **2013**, *53*, 346–354. [CrossRef]

48. Xu, M.; Jin, Z.; Simsek, S.; Hall, C.; Rao, J.; Chen, B. Effect of germination on the chemical composition, thermal, pasting, and moisture sorption properties of flours from chickpea, lentil, and yellow pea. *Food Chem.* **2019**, *295*, 579–587. [CrossRef]

49. Tsatsaragkou, K.; Yiannopoulos, S.; Kontogiorgi, A.; Poulli, E.; Krokida, M.; Mandala, I. Effect of Carob Flour Addition on the Rheological Properties of Gluten-Free Breads. *Food Bioprocess Tech.* **2014**, *7*, 868–876. [CrossRef]

50. Miñarro, B.; Albanell, E.; Aguilar, N.; Guamis, B.; Capellas, M. Effect of legume flours on baking characteristics of gluten-free bread. *J. Cereal Sci.* **2012**, *56*, 476–481. [CrossRef]

51. Rocha Parra, A.F.; Ribotta, P.D.; Ferrero, C. Apple pomace in gluten-free formulations: Effect on rheology and product quality. *Int. J. Food Sci. Technol.* **2015**, *50*, 682–690. [CrossRef]

52. Tsatsaragkou, K.; Styliani, P.; Ioanna, M. Structural role of fibre addition to increase knowledge of non-gluten bread. *J. Cereal Sci.* **2016**, *67*, 58–67. [CrossRef]

53. Marchetti, L.; Cardós, M.; Campaña, L.; Ferrero, C. Effect of glutens of different quality on dough characteristics and breadmaking performance. *LWT* **2012**, *46*, 224–231. [CrossRef]

54. Delta E 101. Available online: http://zschuessler.github.io/DeltaE/learn/ (accessed on 13 January 2020).

55. Fratelli, C.; Muniz, D.G.; Santos, F.G.; Capriles, V.D. Modelling the effects of psyllium and water in gluten-free bread: An approach to improve the bread quality and glycemic response. *J. Funct. Foods* **2018**, *42*, 339–345. [CrossRef]

56. Devries, J.; Camire, M.; Cho, S.; Craig, S.; Gordon, D.; Jones, J.M.; Li, B.; Lineback, D.; Prosky, L.; Tungland, B. The definition of dietary fiber. *Cereal Food World* **2001**, *46*, 112–129.

57.   Melini, V.; Melini, F. Gluten-Free Diet: Gaps and Needs for a Healthier Diet. *Nutrients* **2019**, *11*, 170. [CrossRef]

58.   Vici, G.; Belli, L.; Biondi, M.; Polzonetti, V. Gluten free diet and nutrient deficiencies: A review. *Clin. Nutr.* **2016**, *35*, 1236–1241. [CrossRef]

 *foods* 

Article

Yogurt as an Alternative Ingredient to Improve the Functional and Nutritional Properties of Gluten-Free Breads

Carla Graça, Anabela Raymundo and Isabel Sousa *

LEAF—Linking Landscape, Environment, Agriculture and Food, Research Center of Instituto Superior de Agronomia, Universidade de Lisboa, Tapada da Ajuda, 1349-017 Lisboa, Portugal; carlalopesgraca@isa.ulisboa.pt (C.G.); anabraymundo@isa.ulisboa.pt (A.R.)
* Correspondence: isabelsousa@isa.ulisboa.pt; Tel.: +351-21-365-3246

Received: 25 November 2019; Accepted: 19 January 2020; Published: 21 January 2020

Abstract: Absence of gluten in bakery goods is a technological challenge, generating gluten-free breads with low functional and nutritional properties. However, these issues can be minimized using new protein sources, by the addition of nutritional added-value products. Fresh yogurt represents an interesting approach since it is a source of protein, polysaccharides, and minerals, with potential to mimic the gluten network, while improving the nutritional value of gluten-free products. In the present work, different levels of yogurt addition (5% up to 20% *weight/weight*) were incorporated into gluten-free bread formulations, and the impact on dough rheology properties and bread quality parameters were assessed. Linear correlations ($R^2 > 0.9041$) between steady shear (viscosity) and oscillatory (elastic modulus, at 1 Hz) values of the dough rheology with bread quality parameters (volume and firmness) were obtained. Results confirmed that the yogurt addition led to a significant improvement on bread quality properties, increasing the volume and crumb softness and lowering the staling rate, with a good nutritional contribution in terms of proteins and minerals, to improve the daily diet of celiac people.

Keywords: gluten-free bread; yogurt; rheology

---

1. Introduction

Celiac disease is an immune enteropathy caused by the ingestion of gluten in genetically susceptible individuals, and it is estimated to affect about 1–3% of the population worldwide [1].

Currently, the gluten-free (GF) diet is the only option for people suffering from gluten-related disorders. For other reasons, a high number of nonceliac individuals are also adopting this diet. Subsequently, there has been a significant increase of research work to improve the functional and nutritional properties of GF products [2,3]. However, several studies stated that the consumers remain unsatisfied with the quality of the GF products in the market [4], highlighted their insufficiency on overall appearance and nutritional values, compared to the gluten-containing counterparts [5,6].

Gluten-free products, especially breads being mainly based on refined flours and starches, are generally characterized by poor technological quality attributes, including dry, crumbling texture, color, and mouth feel [7], and undergo fast staling [8].

In terms of nutritional profile, normally they present deficient values of protein and minerals and higher carbohydrates and fat content than recommended [9].

Rice flour has been widely proposed as an alternative for making gluten-free breads (GFB) due to its hypoallergenic protein, soft taste, and white color [10]. However, rice-based bread presented low quality attributes, in terms of volume and hard crumb [11]. Alternative flours from pseudocereal sources, such as amaranth, buckwheat, quinoa, sorghum, and teff, have been applied to improve the nutritional profile of the GFB [12].

In fact, gluten possesses unique viscoelastic properties which are crucial for the water-holding capacity of the dough and gas retention during fermentation [13]. To mimic this structure-building potential, hydrocolloids such as hydroxy-propril-methyl-cellulose, carboxy- or methyl-cellulose, locust bean, guar gum, and xanthan gum, as well as enzymes [8,14], are currently used to improve the viscoelastic properties and technological quality of the end-products [15]. Although these functional additives contribute to the gas retention during the fermentation process, improving the bread volume, they can be insufficient in terms of desirable bread texture and nutritional value.

Addition of new protein sources, e.g., dairy dry powders incorporation (skim milk powder, dry milk, whey protein concentrate) was shown to improve the volume, appearance, and sensory aspects of the loaves [16,17]. A previous work [18] showed a positive impact of dairy products addition, such as fresh yogurt, on gluten-containing bread where a significant improvement on functional properties and nutritional value was noticeable.

Yogurt (Yg), is considered the most popular dairy product worldwide for its nutritional and health benefits, since it is a source of protein (casein), exopolysaccharides (EPS), vitamins (B2, B6, and B12), and minerals (such as Ca, P, and K), representing an interesting alternative for new bakery products [18,19]. In this context, incorporation of fresh Yg in gluten-free bread formulations can be an interesting approach to mimic the gluten network, while improving the nutritional value of gluten-free breads.

The aim of the present work was to explore the potential of fresh yogurt addition on GF bread making and to overcome the technological challenge involved in gluten removal, improving the GFB quality. Different levels of Yg addition to a GF dough formulation, previously optimized, was tested, and the impact on dough, based on steady shear behavior and oscillatory measurements, was assessed. The effect on GFB quality, by the evolution of the crumb firmness and staling rate, during the storage time, as well as from bread quality parameters and nutritional profile, was also evaluated.

2. Materials and Methods

2.1. Raw Materials

Gluten-free bread was prepared using rice flour (composition per 100 g: Moisture content 10.9 g, protein 7.0 g, lipid content 1.3 g, carbohydrates 80.0 g, fiber 0.5 g), buckwheat flour (composition per 100 g: Moisture 13.1 g, protein 13.3 g, lipid content 3.4 g, carbohydrates 61.5 g, fiber 10.0 g, salt 0.08 g, phosphorus 0.35 g, and magnesium 0.23 g), and potato starch (composition per 100 g: Moisture content 17.5 g, protein 0.2 g, lipid content 0.1 g, carbohydrates 80.0 g).

The fresh yoghurt (Yg) used was a commercial product from LongaVida, Portugal (composition per 100 g: Moisture content 88.5 g, protein 3.7 g, lipid content 3.7 g, carbohydrates 5.5 g, fiber 0.8 g, 0.06 g of salt, and 0.12 g of calcium). The dry extract of Yg was determined from the standard Portuguese method: NP.703-1982 (Standard Portuguese Norm), corresponding to 11.5% of dry matter.

Commercial white crystalline saccharose (Sidul, Santa Iria de Azóia, Portugal), sea salt (Vatel, Alverca, Portugal), baker's dry yeast (Fermipan, Lallemand Iberia, SA, Setúbal, Portugal), vegetable oil (Vegê, Sovena Group, Algês, Portugal), and xanthan gum (Naturefoods, Lisboa, Portugal) were also used.

2.1.1. Bread Dough's Preparation and Bread making

Gluten-free bread dough was prepared in a thermo-processor (Bimby-Vorwerk, Cloyes-sur-le-Loir, France), according to the procedure earlier described [20], with some modifications. First, the yeast was activated in warm water on the processor cup, during 2 min at position 3. The rest of the ingredients were added and homogenized during 1 min at position 6, and the kneading was carried out during 10 min. Hydroxy-propril-methyl-cellulose (HPMC) was replaced by xanthan gum and the fermentation time was reduced to 20 min. Preliminary assays showed that 20 min of fermentation was enough to obtain expanded doughs ready to bake. Further fermentation time led to dough structure breakdown, losing the gas (CO_2) produced and retained during this bread making step. Bread baking conditions: Oven with convection at 180 °C during 30 min.

Different GFB formulations tested are summarized in Table 1. Other ingredients added were kept constant: Salt, 1.5%; sugar, 2.8%; dry yeast, 2.8%; xanthan gum, 0.5%; and vegetable oil, 5.5%.

Table 1. Gluten-free bread formulations of control bread (CB) and breads obtained with different levels of yogurt (Yg) addition (YgB), considering the dry extract coming from Yg (11.5%) to replace on flour basis.

Ingredients	CB	YgB$_{5\%}$	YgB$_{10\%}$	YgB$_{15\%}$	YgB$_{20\%}$
Buckwheat	16.6	16.4	15.3	14.4	13.3
Rice	24.8	24.6	23.0	21.5	19.9
Potato starch	13.8	13.7	12.8	12.0	11.1
Yoghurt	0.0	5.0	10.0	15.0	20.0
Added water *	37.5	32.6	31.7	30.7	31.1
Yoghurt water **	0.0	4.7	8.5	11.6	14.1
Total water absorption ***	37.5	37.3	40.3	42.3	45.2

* Determined by mixing curves performed in the MicrodoughLab equipment; ** water coming from Yg addition; *** sum of water added and water coming from Yg addition.

Fixing the viscosity torque values of the gluten-free control dough (16.0 ± 2.0 milliNewton meter - mNm), previously optimized, the water flour absorption was determined for each Yg bread formulation tested (based on 14% of moisture basis), considering the water coming from Yg additions, applying the mixing curves procedure by MicrodoughLab assays (Perten Instruments, Hägersten, Sweden). Control dough and doughs obtained with Yg incorporations, were prepared considering the levels 10, 20, 30, and 40 g of Yg into dough, corresponding to 5% up to 20% *w/w* (weight/weight) in overall percentage. Replacements were based on gluten-free flours basis, substituting the dry extract of each Yg percentage on 100 g of flour [18].

2.1.2. Dough Rheology Measurements

All rheological measurements were carried out in a controlled stress rheometer (Haake Mars III—Thermo Scientific, Karlsruhe, Germany), with a universal temperature control Peltier system to control temperature, using a serrated parallel plate sensor system (PP35 and 1-mm gap), to overcome the slip effect [21,22].

The rheology features of the dough were evaluated after 20 min of fermentation time, and all the assays were conducted at 5 °C of temperature to inactivate the yeast fermentative activity.

The impact of the yogurt addition on dough viscosity behavior was assessed by flow curves, under shear steady conditions ranging the shear rate from 1.0×10^{-6} to 1.0×10^3 s^{-1}.

The Carreau model was used to model the flow curves obtained, applying the Equation (1):

$$\eta = \eta_0/[1 + (\gamma'/\gamma'_c)^2]^{\,s} \tag{1}$$

where η is the apparent viscosity (Pa s), γ is the shear rate (s−1), η_0 is the zero-shear rate viscosity (Pa s), γ'_c is a critical shear rate for the onset of the shear-thinning behavior (s−1), i.e., the value corresponding to the transition from Newtonian to shear-thinning behavior, and s is a dimensionless parameter related to the slope of this region.

Frequency sweep was applied to evaluate the impact of the Yg addition on dough structure, and the evolution of the viscoelastic functions, storage (G′) and loss (G″) moduli, were obtained ranging the frequency from 0.001 Hz to 100.0 Hz, at a constant shear stress (10 Pa), within the linear viscoelastic region of each sample, previously determined (at 1 Hz).

All rheology determinations were repeated at least three times to ensure the reproducibility of the results.

2.2. Quality Assessment of the Gluten-Free Breads

2.2.1. Bread Firmness and Staling Rate

Bread texture was evaluated using a texturometer TA-XTplus (Stable MicroSystems, Surrey, UK) in penetration mode, according to the method earlier described [18,23].

Comparison of the bread texture, with different contents of Yg, was performed in terms of firmness, and the staling rate of the breads was evaluated, measuring the firmness during a storage time, during 96 h (4 days).

Staling bread rate was described as a function of the Yg incorporation by a linear Equation (2)

$$Firmness = A \times time + B, \tag{2}$$

where A can be considered the staling rate and B the initial firmness of the bread.

2.2.2. Quality Parameters of the Gluten-Free Bread

Resulting gluten-free breads were evaluated based on after baking quality parameters, such as moisture, water activity (aw), bake loss (BL), and specific bread volume (SBV) (cm^3/g), as earlier described [18].

Bread moisture was determined according to the standard method (American Association of Cereal Chemists (AACC 44–15.02)). Water activity (aw) variations were determined at room temperature (Hygrolab, Rotronic, Bassersdorf, Switzerland). Bread volume was measured by the rapeseed displacement standard method AACC 10-05.01, after two hours of bread cooling down. Specific volume (cm^3/g) was calculated as the ratio between the volume of the bread and its weight. Weight loss during baking (baking loss) was assessed by weighing the bread forms before and after baking. These measurements were carried out in triplicates.

Bread crumb color was recorded using a Minolta colorimeter (Chromameter CR-300, Minolta—Osaka, Japan) after calibration with a white calibration plate ($L^* = 97.21$, $a^* = -0.14$, $b^* = 1.99$). The data collected from three slices of each bread measured at three different locations of the slices were averaged and expressed using illuminative D65 by L^* a^* b^* scale, where: L^* indicates lightness, a^* indicates hue on a green (−) to red (+) axis, and b^* indicates hue on a blue (−) to yellow (+) axis.

2.3. Nutritional Composition of the Gluten-Free Breads

Nutritional profile characterization of the gluten-free breads was based on protein (International Organization for Standardization (ISO-20483:2006)), lipids (NP 4168), ash (AACC Method 08-01.01), carbohydrates (calculated by difference), and total minerals contents (ICP-AES-Inductively Coupled Plasma-Atomic Emission Spectrometry: Thermo System, ICAP-7000 series) as described in a previous study [18]. All the experiments were performed in triplicate.

2.4. Statistical Analysis

The experimental data were statistically analyzed by determining the average values and standard deviation, and the significance level was set at 95% for each parameter evaluated. Statistical analysis (RStudio, Version 1.1.423, Northern Ave, Boston) was performed by applying variance analysis, the one factor (ANOVA), and post hoc comparisons (Tukey test). Experimental rheology data was fitted to nonlinear Carreau model, using the TA Instruments' TRIOS software.

3. Results and Discussion

3.1. Dough Rheology Measurements

3.1.1. Steady Shear Flow Curves

The effect of the yogurt addition, at different levels, on steady shear behavior of the gluten-free doughs was evaluated, and the flow curves obtained are presented in Figure 1. It can be observed that all systems showed a typical shear-thinning behavior: An initial Newtonian region with constant viscosity at low shear rate, and as the shear rate values increase the dough viscosity began to decrease, following a straight-line decay. Similar results were obtained from the study of the different hydrocolloids' (e.g., xanthan gum) and dairy proteins' interaction on rheology properties of GFB formulations [17].

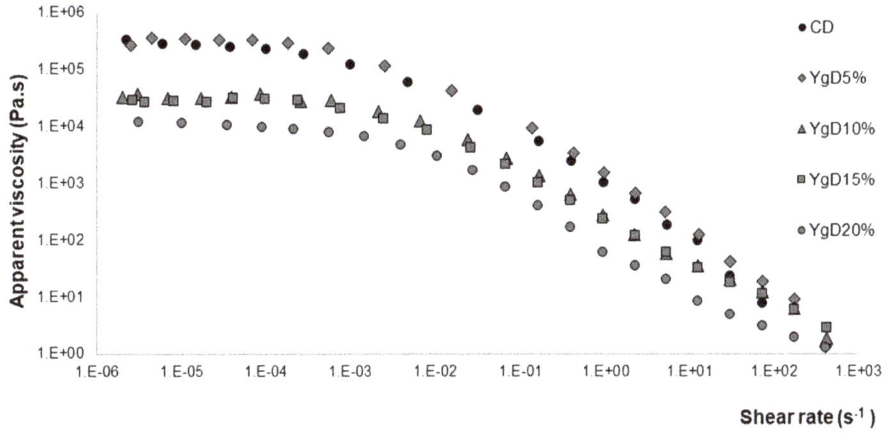

Figure 1. Flow curves, under steady shear conditions, obtained for control dough (CD) and doughs obtained with different levels of Yg addition (YgD).

The experimental data presented in Figure 1 were fitted well by the Carreau model ($R^2 > 0.967$), and the values of the main parameters that characterize the flow behavior are summarized in Table 2.

Table 2. Carreau model parameters obtained for control dough (CD) and doughs obtained with different levels of Yg addition (YgD) *.

Yg levels (%)	η_0 (k Pa s)	$\dot{\gamma}_C$ (s^{-1})	s (slope)	R^2
CD	290.00 ± 8.4 [a]	$1.70 \times 10^{-3} \pm 2.50 \times 10^{-4}$ [a]	0.27 ± 0.03 [a]	0.988
YgD$_{5\%}$	295.04 ± 13.4 [a]	$2.20 \times 10^{3} \pm 1.41 \times 10^{-4}$ [a]	0.23 ± 0.03 [a]	0.980
YgD$_{10\%}$	41.90 ± 1.8 [b]	$2.52 \times 10^{-3} \pm 1.34 \times 10^{-4}$ [a]	0.19 ± 0.03 [ab]	0.977
YgD$_{15\%}$	28.70 ± 0.8 [c]	$2.40 \times 10^{-3} \pm 1.70 \times 10^{-4}$ [a]	0.19 ± 0.04 [ab]	0.976
YgD$_{20\%}$	9.50 ± 0.3 [d]	$1.43 \times 10^{-3} \pm 1.80 \times 10^{-4}$ [a]	0.17 ± 0.06 [b]	0.967

* Different letters (a, b, c, d) within the same column indicate significant statistical differences at $p \le 0.05$ (Tukey test) compared with the control bread values.

As it can be observed from the zero-shear rate viscosity values (η_0), higher dough viscosities were obtained for the control dough (CD) and yoghurt dough at 5% (YgD$_{5\%}$, lower level of Yg tested). According to previous works [17,24], GF doughs obtained with xanthan gum (XG) addition showed high viscosities and flow behavior indexes due to the complex aggregates formed by strong molecular linkages. However, a balance must be reached since high viscosity may retain bubbles in the batter, but it may also restrict expansion during baking [25].

Nevertheless, increasing the amounts of Yg resulted in a significant decrease of dough viscosity (η_0), varying from 290.00 k Pa s (CD) to 9.50 k Pa s ($YgD_{20\%}$), representing a reduction of around 96.60%. One can suggest that the Yg incorporation promoted a dilution effect of starch–cereal protein–xanthan gum interaction density, decreasing the dough viscosity under shear conditions. In addition, the presence of the casein and of the exopolysaccharides (EPS) coming from Yg (produced by lactic acid bacteria) impacted the system, improving the lubrication and flexibility of the dough network. These effects can also be explained by the ability of EPS to bind water and retain moisture, contributing to the increase in the water-holding capacity [26] and, possibly, reducing the rigidity of the linkages between the different molecules of the GF dough matrix. Similar findings were obtained by other authors [27], investigating the addition of caseins and albumins on GFB formulations, where a considerable reduction on dough viscosity was obtained.

In terms of critical shear rate, no significant differences were observed. All the dough systems' viscosity began to decrease around 2.0×10^{-3} s^{-1}. This behavior suggests that, although the addition of Yg can promote the dilution effect on molecular links density as well as reducing the stiffer complex aggregates, it seems to have no effect on the breakdown of the dough matrix.

3.1.2. Dough Viscoelastic Behavior

The changes on viscoelastic behavior, expressed in terms of elastic (G′) and viscous (G″) moduli, of the gluten-free doughs, with different levels of Yg addition, were evaluated on fermented doughs, by oscillatory frequency sweep measurements.

The comparison of the mechanical spectra of control dough (CD) and doughs obtained with Yg additions (YgD) are represented in Figure 2.

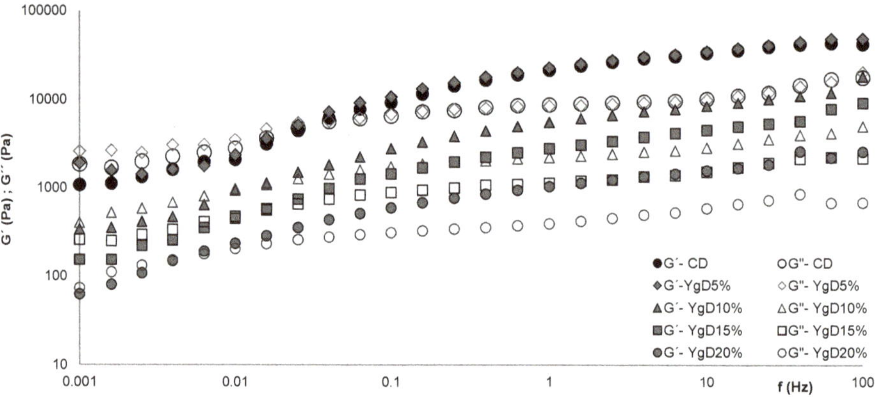

Figure 2. Changes in viscoelastic functions, elastic (G′) and viscous (G″) moduli, promoted by different levels of Yg addition, $YgD_{5\%}$ up to $YgD_{20\%}$, compared to control dough (CD).

Figure 2 shows that the elastic (G′) and viscous (G″) moduli values obtained for Yg doughs, from 10% of Yg addition, were lower than control dough (CD), indicating a formation of weaker structures more like batter. No significant differences in viscoelastic profile for lower values of Yg tested ($YgD_{5\%}$) were observed, comparing with CD.

These results are aligned with those obtained by steady shear flow curves, discussed above, where a significant reduction of dough viscosity was registered from 10% on of Yg addition. Similar results were obtained by other researchers [10,28], using different gluten-free flours, proteins sources, and hydrocolloids in GFB formulations.

Analyzing in detail the results (Figure 2) at low frequencies, all the dough systems displayed a viscoelastic fluid behavior with values of G″ higher than G′, characteristic of an entangled network [29–31], probably formed by the proteins', exopolysaccharides', and starch molecules' interaction. However, with the frequency increase, the crossover of both moduli occurred and the dominance of the G′ over the G″ was observed, expressing a typical pseudo-gel behavior [28–30], with high frequency dependence [32]. These results agree with previous findings obtained by the study of the rheology evolution of gluten-free flours and starches during bread fermentation and baking [33].

It can be observed that the frequency values of the G′ crossing over the G″ were reduced by the Yg addition to dough, varying from 0.025 Hz for CD to 0.004 Hz for $YgD_{20\%}$ (higher level tested). Based on these results, and although the Yg addition promoted significant changes on dough structure, some reinforcement of the molecular linkages on the dough matrix should be considered. This may be explained by the presence of casein and exopolysaccharides coming from Yg that are acting on the system through three probable mechanisms:

(1)   Improving the orientation and disentanglement of the molecular linkages on the dough matrix, under oscillatory conditions [34], by lubrification and flexiblizing effects,
(2)   Reducing the stiffer network structure formed by xanthan gum, giving more flexibility to the dough network [17], and
(3)   Reinforcing the dough molecular bonds, by additional casein and exopolysaccharides interaction, giving more stability to the dough [13,17].

These results showed that the Yg additions promoted considerable changings in viscoelastic properties that resulted in the improvement of the dough network capacity to incorporate and retain gas bubbles produced by yeast's fermentative activity, consequently, reducing significantly the viscoelastic properties of the dough under oscillatory conditions. This observed behavior resulted in better texture properties and higher specific volumes of the bread [13].

From previous work, it was already stated that interactions between protein and polysaccharides led to similar changes in the viscoelastic functions (G′ and G″) profile of gluten-free bread doughs [10]. These findings are also in line with those obtained by other researchers [17] evaluating the effect of different hydrocolloids and proteins on rheology properties of GFB formulation.

3.2. Evaluation of the Gluten-Free Bread Properties

3.2.1. Bread Texture and Staling Rate

The texture of the GFB produced with different contents of Yg addition was evaluated based on bread crumb firmness by a puncture test, and subsequent bread staling rate obtained from the evolution of the bread firmness, during the storage time of 96 h at room temperature [18,23].

From Figure 3, it can be observed that the Yg addition had a significant ($p \leq 0.05$) positive impact to increase the bread crumb softness: Initial and final values of control bread firmness were higher than all levels of Yg tested, varying from 2.25 Newton (initial) to 6.50 Newton (final) for CB and 0.82 Newton (initial) to 1.35 Newton (final) for higher levels of Yg tested on breads ($YgB_{20\%}$), representing a decrease of 64% and 80% of crumb firmness values, respectively.

Bread staling rate was described as a function of time ($R^2 > 0.974$). The linear parameters presented in Table 3 clearly reflect the impact of Yg additions on bread staling rate (A, the slope) and initial firmness (B, the interception).

The staling rate of the Yg breads were significantly lower than for the (CB), ranging from 0.044 N/h (CB) to 0.005 N/h for higher level of yoghurt tested yoghurt bread ($YgB_{20\%}$), representing a reduction of 90%. Similar results were obtained by other authors [35] evaluating the impact of the dairy powders on loaf and crumb characteristics and on shelf life of GFB. These findings are also in line with those obtained by a previous work [18] evaluating the impact of fresh dairy products' addition on technological, nutritional, and sensory properties of wheat bread. Good results of bread texture and staling rate can be explained by the presence of exopolysaccharides, coming from the Yg addition, since it has been well demonstrated that EPS improve bread texture properties, and such effects are related to its ability to bind water and retain moisture, retarding the starch crystallization and, hence, the increase of bread firmness [26,36–38].

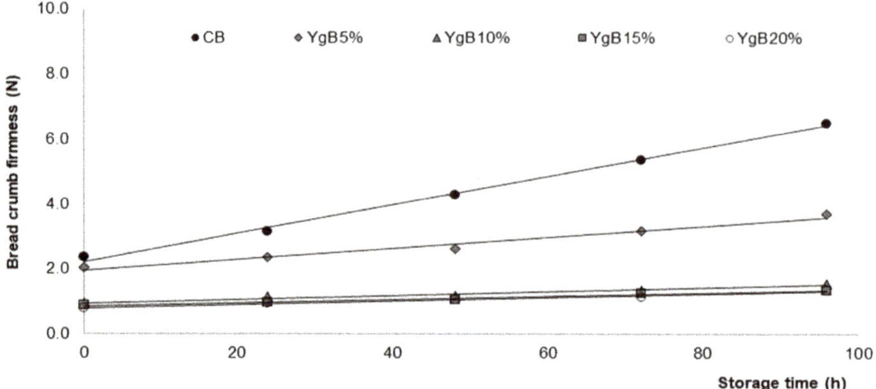

Figure 3. Variation of bread crumb firmness during 96 h of storage time, at room temperature, obtained for breads with different levels of Yg addition (YgB$_{5\%}$ up to YgB$_{20\%}$) compared to control bread (CB).

Table 3. Bread stalling parameters: A, bread staling rate (Newton/h), and B, initial bread firmness (N), obtained for control bread (CB) and breads with different levels of Yg tested (YgB) *.

Yg Levels	B—Initial Firmness (N)	A—Staling Rate (N/h)	R^2
CB	2.25 [a]	0.044 [a]	0.997
YgB$_{5\%}$	1.90 [b]	0.020 [b]	0.974
YgB$_{10\%}$	0.95 [c]	0.006 [c]	0.975
YgB$_{15\%}$	0.85 [c]	0.005 [c]	0.979
YgB$_{20\%}$	0.82 [c]	0.005 [c]	0.993

* Different letters (a, b, c) within the same column indicate significant statistical differences at $p \leq 0.05$, (Tukey test), compared with the control bread parameters.

It can be stated that the Yg addition increased the bread crumb softness and delayed the staling rate of the GFB, leading to an increase of shelf life, which is an important industrial advantage.

3.2.2. Quality Parameters of Gluten-Free Bread

The impact of the Yg addition on gluten-free quality parameters, such as crumb color, moisture, water activity (aw), bake loss (%), and specific bread volume (SBV), was evaluated. Results obtained are summarized in Table 4.

Table 4. Gluten-free bread quality parameters: Crumb color (L*, a*, b*), moisture, water activity (aw), bake loss (BL), and specific bread volume (SBV) of the control bread (CB) and breads produced with different levels of Yg (YgB) addition *.

Samples	L*	a*	b*	Moisture (%)	aw	BL (%)	SBV (cm^3/g)
CB	59.52 ± 2.85 [a]	7.81 ± 0.23 [a]	11.40 ± 0.20 [a]	42.00 ± 0.33 [a]	0.959 ± 0.006 [a]	10.00 ± 0.41 [a]	1.80 ± 0.08 [a]
YgB$_{5\%}$	59.93 ± 2.83 [a]	6.21 ± 0.08 [ab]	13.61 ± 1.20 [a]	43.20 ± 0.11 [a]	0.979 ± 0.001 [a]	9.35 ± 0.93 [a]	2.10 ± 0.03 [b]
YgB$_{10\%}$	62.52 ± 2.75 [ab]	5.26 ± 0.17 [b]	19.60 ± 0.75 [b]	46.40 ± 0.20 [b]	0.978 ± 0.002 [a]	8.44 ±1.24 [ab]	2.40 ± 0.05 [c]
YgB$_{15\%}$	68.02 ± 1.60 [b]	4.16 ± 0.11 [c]	22.90 ± 0.50 [bc]	48.00 ± 0.04 [c]	0.981 ± 0.005 [a]	7.50 ± 0.30 [b]	2.43 ± 0.08 [c]
YgB$_{20\%}$	68.60 ± 2.65 [b]	4.05 ± 0.15 [c]	25.32 ± 0.73 [c]	50.00 ± 0.83 [d]	0.985 ± 0.003 [a]	7.30 ± 0.94 [b]	2.50 ± 0.06 [d]

* Different letters (a, b, c, d) within the same column, for each level of Yg, indicate statistically significant differences at $p \leq 0.05$ (Tukey test), compared with the control bread parameters.

Bread crumb color was significantly changed by Yg addition, giving a lighter color (higher L* values) with more pronounced yellow tone (higher b* values) while the red tone became less dominant (lower a* values).

In terms of bread moisture, an increase in moisture content was observed (from 10% on, of Yg addition), varying from 42.0 for CB to 50.0 for $YgB_{20\%}$ (higher level tested), corresponding to an increase of 20%. Related to water activity, no significant ($p \geq 0.05$) differences were registered for all levels tested, compared to CB.

The bake loss (BL) represented the amount of water and organic material (CO_2 and other volatiles) lost during baking [12]. One can see (Table 4) that increasing the amounts of Yg led to a significant decrease of BL ($p \leq 0.05$), corresponding to an increase of 27% on bread yield, comparing CB with $YgB_{20\%}$. Presumably, as discussed above, the dough network formed by the Yg addition, probably via starch molecules, vegetable protein, casein, and EPS (coming from Yg), had a higher ability to trap water and it may cause an increase in water-holding capacity, consequently, a decrease on the BL percentage [39–41]. These results agree with those published by other authors [26] on the combination of dairy proteins with transglutaminases on GFB formulation.

Based on specific bread volume (SBV), a significant improvement was observed by Yg addition from 10% on, varying between 1.80 cm^3/g for CB to 2.50 cm^3/g for $YgB_{20\%}$, representing an increase of around 40%. These improvements on the SBV were probably due to the increase in water-holding capacity, contributing to the starch molecules being more prone to form a uniform continuous starch–protein matrix, which was further enhanced during baking [12,41]. In addition, the contribution of the caseins and exopolysaccharides, coming from Yg addition, cannot be excluded, probably by building a structured starch–protein–EPS matrix [38,39], improving the flexibility of the dough network and the capacity to retain the gases produced during the fermentation process, contributing to keeping the bread structure with uniform gas cell distribution, resulting in better bread volumes than CB, as can be observed in Figure 4.

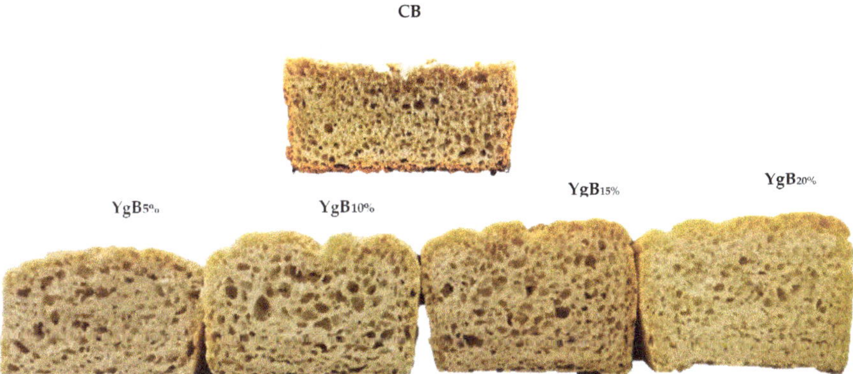

Figure 4. Control bread (CB) and breads obtained with different levels of Yg addition (YgB): 5, 10, 15, and 20% (*w/w- weight/weight*).

Our results agree with those published by other authors [26] reporting the effect of dairy proteins (caseins and albumins) with transglutaminase additions on the improvement of crumb texture and better SBV.

Resuming, it can be stated that the Yg addition improved the bread quality of the gluten-free bread in all baking quality parameters evaluated.

3.3. Relationship between Bread Quality Parameters and Dough Rheology Properties

Bread crumb texture and volume are considered the most important baking properties by the consumers [12] and the incorporation of fresh Yg showed to be a potential ingredient to improve these bread quality parameters.

The results presented along this work suggested a relationship between bread quality parameters and dough rheology properties. Linear correlations ($R^2 > 0.9041$) between the bread firmness (BF) and specific bread volume (SBV) with steady shear (dough viscosity (DV)) and oscillatory (G, elastic modulus at 1Hz) values of dough rheology were obtained. The linear correlations are illustrated in Figure 5.

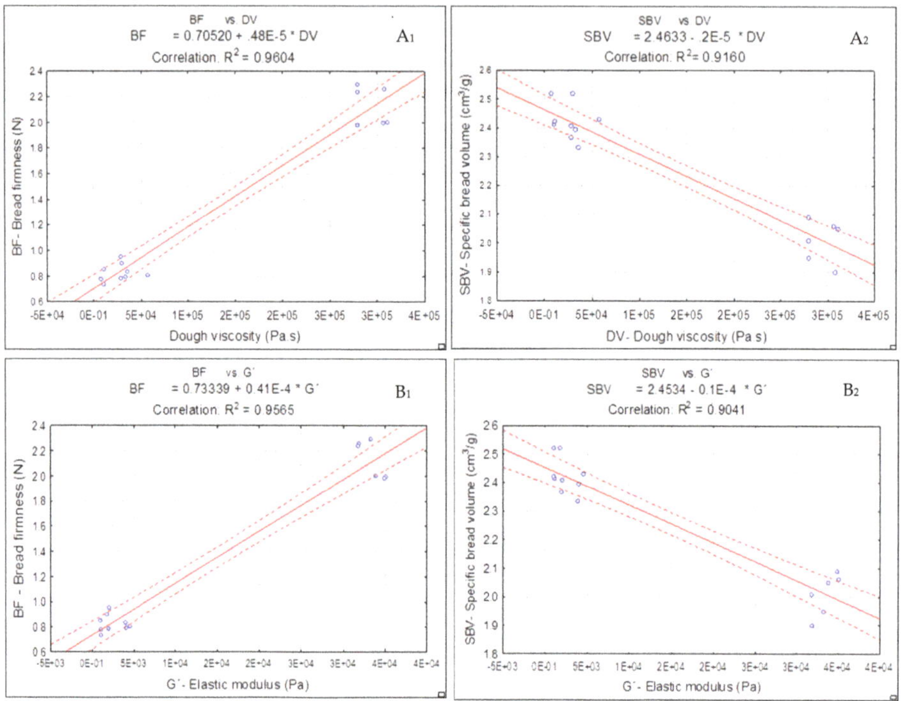

Figure 5. Linear correlation between bread firmness (BF) and specific bread volume (SBV) with dough viscosity (DV) and elastic modulus at 1 Hz (G′): (A_1) BF vs. DV, (A_2) SBV vs. DV, (B_1) BF vs. G′, (B_2) SBV vs. G′.

As can be observed from Figure 5, while the control dough (only with xanthan gum) exhibited higher dough viscosity (DV) and elasticity (G′ at 1Hz) values, the specific volume of this bread was not high, and the crumb firmness was harder than the Yg doughs obtained. Previous works [13,17] reported that, although xanthan gum had the most pronounced effect on viscoelastic properties of the dough, the bread texture and volume were negatively affected.

In opposite, lower values of dough viscosity (DV) and elastic modulus (G′ at 1 Hz) by the Yg addition to dough resulted in breads with lower values of crumb firmness (B1) and staling rate and with higher specific volumes (B2) than control bread.

These linear correlations support the results presented along this work, showing a strong correlation between bread quality parameters and dough rheology properties. It can be stated that the dough system was improved by Yg addition, which resulted in softer breads with better volumes, as aforementioned.

These findings disagree with those obtained by other researchers [11,17] evaluating different gums and emulsifiers on gluten-free bread formulations, where higher dough viscosity and elastic values resulted in lower firmness values of the bread. It suggests that the quality of the GFB depends strongly on the type of the ingredients used and the interactions between the macromolecules into play, to mimic the structure building-like gluten matrix.

These relations between the dough rheology and bread quality can be useful to predict the behavior of the dough and provide important information for the GF bread making industry.

3.4. Nutritional Composition of the Gluten-Free Breads

The nutritional composition of the gluten-free breads, including the mineral profile, were determined for control bread (CB) and breads obtained with 10% and 20% of Yg addition (YgB). A significant ($p \leq 0.05$) reinforcement on protein and ash content was observed for both levels of Yg tested. However, a remarkable effect was obtained for 20% *(w/w)* of Yg addition in both cases, representing an increase of 52% and 55%, respectively. Nutritional composition and minerals profile of the gluten-free breads are summarized in Table 5.

Table 5. Proximate nutritional composition and mineral profile for control bread (CB) and breads enriched with 10% (YgB$_{10\%}$) and 20% (YgB$_{20\%}$) of yogurt *.

g/100 g	CB	YgB10%	YgB20%	
Proteins	5.34 ± 0.21 [a]	6.90 ± 0.02 [b]	8.10 ± 0.20 [c]	
Lipids	4.83 ± 0.29 [a]	5.20 ± 0.84 [a]	5.60 ± 0.22 [a]	
Ash	1.40 ± 0.03 [a]	1.74 ± 0.17 [ab]	2.12 ± 0.27 [b]	
Carbohydrates	45.61 ± 1.12 [a]	39.00 ± 2.25 [b]	34.20 ± 1.58 [c]	
Kcal	250.30 ± 1.40 [a]	237.07 ± 2.13 [b]	226.52 ± 2.89 [c]	
	Minerals content mg/100 g			15% RDV (mg/100g) **
Na (g/100 g)	0.41 ± 0.03 [a]	0.42 ± 0.01 [a]	0.53 ± 0.02 [b]	
K	412.0 ± 8.02 [a]	497.30 ± 5.92 [b]	515.40 ± 16.40 [b]	300.0
P	131.81 ± 5.40 [a]	133.90 ± 0.90 [a]	141.24 ± 2.00 [b]	120.0
Mg	48.70 ± 1.02 [a]	50.24 ± 0.56 [a]	52.32 ± 0.82 [b]	45.0
Ca	35.93 ± 2.06 [a]	52.20 ± 0.81 [b]	92.60 ± 2.60 [c]	120.0
Fe	6.94 ± 0.12 [a]	4.74 ± 0.67 [b]	3.10 ± 0.05 [c]	2.1
Cu	0.23 ± 0.01 [a]	0.25 ± 0.02 [a]	0.30 ± 0.03 [a]	0.2
Mn	0.51 ± 0.01 [a]	0.60 ± 0.02 [a]	0.70 ± 0.03 [b]	0.3
Zn	1.12 ± 0.01 [a]	1.05 ± 0.08 [a]	0.98 ± 0.01 [b]	2.3

* Different letters used (a, b, c, d) indicate statistically significant differences, within the same row, at $p \leq 0.05$ (Tukey test), compared with the bread control; ** according to the recommended daily values (RDV) established by Regulation (European Community), N° 1924/2006; Directive N° 90/494 (CE).

In terms of lipids, no significant ($p \geq 0.05$) differences were obtained. Based on carbohydrates content, a significant decrease for higher levels of Yg tested (20% *w/w*) was observed, representing a decrease of 25%, due to the dilution effect of the starch content.

Regarding the minerals profile, a significant improvement ($p \leq 0.05$) of major (Ca, K, P, Mg) and trace minerals (Fe, Cu, Mn) was obtained, representing, in general, more than 15% of the recommended daily values (Regulation (European community)), No. 1924/2006; Directive No. 90/494 (CE). Considering the higher level of Yg tested (YgB$_{20\%}$), a significant increase in Ca (158.0%), K (25.1%), P (7.0%), and Mg (4.1%) can be noticed as well as in Fe (55.3%), Cu (30.4), and Mn (37.3%), compared to control bread (Table 5). Similar results were obtained by other authors [18] testing the addition of dairy products on wheat bread.

These results showed that the Yg addition can be used to enhance the nutritional value of the GFB, increasing the amount of protein and minerals profile with a good contribution to reducing the carbohydrates intake.

4. Conclusions

Gluten-free bread formulations, with different levels of yogurt addition, were evaluated using dough rheology measurements and baking quality parameters.

The results of the present work showed that the functionality of gluten-free breads, in terms of bread making performances, quality parameters, and nutritional profile can be successfully improved by the addition of fresh yogurt.

Although, the yogurt incorporation significantly reduced the rheology properties of doughs, such effects resulted in significant improvements in the overall quality of the corresponding breads. Good linear correlations between bread firmness, specific volume with flow behavior, and viscoelastic functions were found, supporting the results obtained.

Resuming, the Yg showed to be a potential ingredient to improve the quality of gluten-free breads, resulting in softer breads with higher volume and lower staling rate, compared to control bread.

Related to the nutritional composition, the addition of Yg revealed to be an attractive ingredient to enhance the nutritional value of GF breads, increasing the protein and mineral contents and reducing the carbohydrates intake, with a good contribution to improve the daily diet of the celiac people.

Author Contributions: C.G. conceived and planned the experiments; performed all samples preparation, analysis, data analysis, and interpretation of the results; and wrote the manuscript. A.R. and I.S. supervised the research work, contributed to the discussion of the data, and revised the manuscript. All authors have read and agreed to the published version of the manuscript.

Funding: This work was funded by the University of Lisbon (Doctoral Grant reference to C10781w) and by the Portuguese Foundation for Science and Technology (FCT) through the research unit of University identification/Agronomic/04129/2013 (LEAF, Research Center—Linking Landscape, Environment, Agriculture, and Food).

Conflicts of Interest: The authors declare no conflicts of interest.

References

1. Rosell, C.M.; Barro, F.; Sousa, C.; Mena, M.C. Cereals for developing gluten-free products and analytical tools for gluten detection. *J. Cereal Sci.* **2014**, *59*, 354–364. [CrossRef]
2. Capriles, V.D.; Arêas, J.A.G. Novel approaches in gluten-free breadmaking: Interface between food science, nutrition, and health. *Compreh. Rev. Food Sci. Food Saf.* **2014**, *13*, 871–890. [CrossRef]
3. Capriles, V.D.; dos Santos, F.G.; Arêas, J.A.G. Gluten-free breadmaking: Improving nutritional and bioactive compounds. *J. Cereal. Sci.* **2016**, *67*, 83–91. [CrossRef]
4. Do Nascimento, A.B.; Fiates, G.M.R.; dos Anjos, A.; Teixeira, E. Gluten-free is not enough – perception and suggestions of celiac consumers. *Int. J. Food Sci. Nutri.* **2014**, *65*, 394–398. [CrossRef] [PubMed]
5. Pellegrini, N.; Agostoni, C. Nutritional aspects of gluten-free products. *J. Sci. Food Agric.* **2015**, *95*, 2380–2385. [CrossRef] [PubMed]
6. Calle, J.; Benavent-Gil, Y.; Rosell, C. Development of gluten free breads from Colocasia esculenta flour blended with hydrocolloids and enzymes. *Food Hydrocoll.* **2020**, *98*, 105243. [CrossRef]
7. Gallagher, E.; Gormley, T.R.; Arendt, E.K. Crust and crumb characteristics of gluten free breads. *J. Food Eng.* **2003**, *56*, 153–161. [CrossRef]
8. Ahlborn, G.J.; Pike, O.A.; Hendrix, S.B.; Hess, W.M.; Huber, C.S. Sensory, mechanical, and microscopic evaluation of staling in low-protein and gluten-free breads. *Cereal Chem.* **2005**, *82*, 328–335. [CrossRef]
9. Segura, M.E.; Rosell, C.M. Chemical composition and starch digestibility of different gluten-free breads. *Plant Foods Hum. Nutr.* **2011**, *66*, 224–230. [CrossRef]
10. Torbica, A.; HadnaCev, M.; Dapcevic, T. Rheological, textural and sensory properties of gluten-free bread formulations based on rice and buckwheat flour. *Food Hydrocoll.* **2010**, *24*, 626–632. [CrossRef]
11. Locke, J.; González, L.; Loubes, M.; Tolaba, M. Optimization of rice bread formulation by mixture design and relationship of bread quality to flour and dough attributes. *LWT-Food Sci. Technol.* **2019**, *113*, 108299. [CrossRef]
12. Alvarez-Jubete, L.; Arendt, E.; Gallagher, E. Nutritive value of pseudocereals and their increasing use as functional gluten-free ingredients. *Trends Food Sci. Technol.* **2010**, *21*, 106–113. [CrossRef]

13. Lazaridou, A.; Duta, D.; Papageorgiou, M.; Belc, N.; Biliaderis, C.G. Effects of hydrocolloids on dough rheology and bread quality parameters in gluten-free formulations. *J. Food Eng.* **2007**, *79*, 1033–1047. [CrossRef]

14. Pruska-Kedzior, A.; Kedzior, Z.; Goracy, M.; Pietrowska, K.; Przybylska, A.; Spychalska, K. Comparison of rheological, fermentative and baking properties of gluten-free dough formulations. *Eur. Food Res. Technol.* **2008**, *227*, 1523–1536. [CrossRef]

15. Masure, H.G.; Fierens, E.; Delcour, J.A. Current and forward-looking experimental approaches in gluten-free bread making research. *J. Cereal Sci.* **2016**, *67*, 92–111. [CrossRef]

16. Ziobro, R.; Witczak, T.; Juszczak, L.; Korus, J. Supplementation of gluten-free bread with non-gluten proteins. Effect on dough rheological properties and bread characteristic. *Food Hydrocoll.* **2013**, *32*, 213–220. [CrossRef]

17. Demirkesen, I.; Mert, B.; Sumnu, G.; Sahin, S. Rheological properties of gluten-free bread formulations. *J. Food Eng.* **2010**, *96*, 295–303. [CrossRef]

18. Graça, C.; Raymundo, A.; Sousa, I. Wheat Bread with Dairy Products—Technology, Nutritional, and Sensory Properties. *Appl. Sci.* **2019**, *9*, 4101. [CrossRef]

19. Sharafi, S.; Yousefi, S.; Faraji, A. Developing an innovative textural structure for semi-volume breads based on interaction of spray-dried yogurt powder and jujube polysaccharide. *Int. J. Biol. Macromol.* **2017**, *104*, 992–1002. [CrossRef]

20. Martins, R.B.; Fernandes, I.; Nunes, M.C.; Barros, A.; Peres, J.; Raymundo, A. Improving gluten free bread quality using by-products and underexploited resources. In *Livro de Resumos do XIV Encontro Luso-Galego de Química*; Sociedade Portuguesa de Química: Porto, Portugal, 2018.

21. Franco, J.M.; Raymundo, A.; Sousa, I.; Gallegos, C. Influence of processing variables on the rheological and textural properties of lupin protein-stabilized emulsions. *J. Agric. Food Chem.* **1998**, *46*, 3109–3115. [CrossRef]

22. Graça, C.; Raymundo, A.; Sousa, I. Rheology changes in oil-in-water emulsions stabilized by a complex system of animal and vegetable proteins induced by thermal processing. *LWT-Food Sci. Technol.* **2016**, *74*, 263–270. [CrossRef]

23. Graça, C.; Fradinho, P.; Sousa, I.; Raymundo, A. Impact of Chlorella vulgaris on the rheology of wheat flour dough and bread texture. *LWT-Food Sci. Technol.* **2018**, *89*, 466–474. [CrossRef]

24. Sworn, G. Xanthan gum. In *Handbook of Hydrocolloids*, 1st ed.; Philips, G.O., Williams, P.A., Eds.; Woodhead: Cambridge, UK, 2000; pp. 103–115.

25. Gómez, M.; Moraleja, A.; Oliete, B.; Ruiz, E.; Caballero, P.A. Effect of fibre size on the quality of fibre-enriched layer cakes. *LWT-Food Sci. Technol.* **2010**, *43*, 33–38. [CrossRef]

26. Lynch, K.M.; Coffey, A.; Arendt, E.K. Exopolysaccharides producing lactic acid bacteria: Their techno-functional role and potential application in gluten-free bread products. *Food Res. Int.* **2017**, *110*, 52–61. [CrossRef]

27. Storck, C.; da Rosa Zavareze, E.; Gularte, M.; Elias, M. Protein enrichment and its effects on gluten-free bread characteristics. *LWT-Food Sci. Technol.* **2013**, *53*, 346–354. [CrossRef]

28. Torbica, A.; Belović, M.; Tomić, J. Novel breads of non-wheat flours. *Food Chem.* **2019**, *282*, 134–140. [CrossRef]

29. Hundschell, C.S.; Wagemans, A.M. Rheology of common uncharged exopolysaccharides for food applications. *Cur. Opin. Food Sci.* **2019**, *27*, 1–7. [CrossRef]

30. Picout, D. Ross-Murphy, Rheology of Biopolymer Solutions and Gels. *Sci. World J.* **2003**, *3*, 105–121. [CrossRef]

31. Marchetti, l.; Andrés, S.; Cerruti, P.; Califano, A. Effect of bacterial nanocellulose addition on the rheological properties of gluten-free muffin batters. *Food Hydrocoll.* **2020**, *98*, 105315. [CrossRef]

32. Razmkhah, S.; Razavi, S.; Mohammadifar, M. Dilute solution, flow behavior, thixotropy and viscoelastic characterization of cress seed (Lepidium sativum) gum fractions. *Food Hydrocoll.* **2017**, *63*, 404–413. [CrossRef]

33. Martinez, M.; Gómez, M. Rheological and microstructural of the most common gluten-free flours and starches during bread fermentation and baking. *J. Food Eng.* **2017**, *197*, 78–86. [CrossRef]

34. Bernaerts, T.; Panozzo, A.; Doumen, V.; Foubert, I.; Gheysen, L.; Goiris, K.; Moldenaers, P.; Hendrickx, M.; Van Loey, A. Microalgal biomass as a (multi) functional ingredient in food products: Rheological properties of microalgal suspensions as affected by mechanical and thermal processing. *Algal Res.* **2017**, *25*, 452–463. [CrossRef]

35. Gallagher, E.; Kundel, A.; Gormley, T.R.; Arendt, E.K. The effect of dairy and rice powder addition on loaf and crumb characteristics, and on shelf life (intermediate and long-term) of gluten-free breads in a modified atmosphere. *Eur. Food Res. Technol.* **2003**, *218*, 44–48. [CrossRef]
36. Galle, S.; Schwab, C.; Dal Bello, F.; Coffey, A.; Gänzle, M.G.; Arendt, E.K. Comparison of the impact of dextran and reuteran on the quality of wheat sourdough bread. *J. Cereal. Sci.* **2012**, *56*, 531–537.
37. Wolter, A.; Hager, A.S.; Zannini, E.; Czerny, M.; Arendt, E.K. Influence of dextran producing *Weissella cibaria* on baking properties and sensory profile of gluten-free and wheat breads. *Int. J. Food Microb.* **2014**, *172*, 83–91. [CrossRef] [PubMed]
38. Arendt, E.K.; Ryan, L.A.; Dal Bello, F. Impact of sourdough on the texture of bread. *Food Microbiol.* **2007**, *24*, 165–174. [CrossRef] [PubMed]
39. Moore, M.M.; Heinbockel, M.; Dockery, P.; Ulmer, H.M.; Arendt, E.K. Network formation in gluten-free bread with application of transglutaminase. *J. Cereal. Chem.* **2006**, *83*, 28–36. [CrossRef]
40. Renzetti, S.; Bello, F.D.; Arendt, E.K. Microstructure, fundamental rheology and baking characteristics of batters and breads from different gluten- free flours treated with a microbial transglutaminase. *J. Cereal. Sci.* **2008**, *48*, 33–45. [CrossRef]
41. Gallagher, E. *Gluten-Free Food Science and Technology*, 1st ed.; Wiley Blackwell: Oxford, UK, 2009; pp. 52–81.

Article

Optimization of the Formula and Processing Factors for Gluten-Free Rice Bread with Tamarind Gum

Ye-Eun Hong [1] and Meera Kweon [1,2,*]

[1] Department of Food Science and Nutrition, Pusan National University, Busan 46241, Korea; yea1244@naver.com

[2] Kimchi Institute, Pusan National University, Busan 46241, Korea

* Correspondence: meera.kweon@pusan.ac.kr; Tel.: +82-51-510-2716; Fax: +82-51-583-3648

Received: 31 December 2019; Accepted: 28 January 2020; Published: 1 February 2020

Abstract: The formula and processing parameters for gluten-free rice bread were optimized using a factorial design, including a center point. Gum concentration (GC), water amount (WA), mixing time (MT), and fermentation time (FT) were selected as factors, and two levels were used for each factor: 1 and 2% for GC; 80 and 100 g for WA; 5 and 10 min for MT; and 30 and 60 min for FT. The WA and FT were identified as the most significant factors in determining the quality of gluten-free rice bread with tamarind gum. Thus, the optimized formula and processing conditions for maximizing bread volume and minimizing bread firmness were 1% gum, 100 g water, 5 min MT, and 60 min FT. The addition of an anti-staling enzyme reduced the increase in bread firmness and the enthalpy of starch retrogradation, suggesting its potential for successful application in commercially manufactured rice bread with tamarind gum.

Keywords: gluten-free; rice bread; tamarind gum; factorial design; optimization; formula; processing factor

---

1. Introduction

In Korea, with increasing personal income and the westernization of dietary patterns, the consumption of rice is decreasing, while the use of wheat is increasing [1]. Wheat contains glutenin and gliadin, which are unique protein components that form the viscoelastic gluten essential for bread-making. However, omega gliadin, present in wheat, is a major component that causes celiac disease, which is an autoimmune disease caused by ingesting gluten, and its main symptoms include chronic diarrhea, vomiting, and abdominal distension in young infants [2]. Patients with celiac disease are advised to consume gluten-free foods from grains that do not contain gliadin. Rice is one such grain commonly used for gluten-free products. Thus, the development of gluten-free foods using rice flour is considered to be a suitable strategy to increase rice consumption [3].

Considerable research on the development of gluten-free foods using rice has been actively conducted worldwide [4–6], and various gluten-free products are being sold commercially. Baked goods are popular, and many studies have been performed to develop gluten-free rice bread. However, the improvement of the pseudo-viscoelasticity imparted by rice flour dough is still needed to develop a gluten-free rice product whose volume and texture are similar to those of bread made of wheat flour. Many types of grains, protein sources, and gums have been studied to improve the quality of gluten-free bread [7,8].

Several gums, such as xanthan gum, guar gum, pectin, carrageenan, and hydroxyl methylcellulose (HPMC), have been studied [9–13], and, recently, tamarind gum has been reported for its suitability in the manufacturing of rice bread [14]. Tamarind gum is a xyloglucan and can improve bread-making by increasing the stability of the dough and loaf volume [15]. The majority of studies on gluten-free products have investigated the ingredients and formulation, with less focus on the processing conditions.

Therefore, in-depth research on the formulation and processing conditions of gluten-free rice bread, with the added tamarind gum, is needed for its successful application.

In this study, a factorial design was used to identify and optimize significant factors in the formulation and processing conditions of gluten-free rice bread with the addition of tamarind gum. Gum concentration (GC), water amount (WA), mixing time (MT), and fermentation time (FT) were the four factors selected, and two levels were set for each factor. Quality characteristics of the gluten-free rice bread were evaluated by measuring the pH of batter before and after fermentation, moisture, volume, and crumb texture of the bread. The effect of an anti-staling enzyme on gluten-free rice bread at the optimal formulation and processing conditions, selected through the experiment, was analyzed for quality changes during cold storage by measuring moisture, texture, and the retrogradation enthalpy of the rice bread by differential scanning calorimetry (DSC).

2. Materials and Methods

2.1. Materials

The rice flour used (Nongsim, Seoul, Korea) was a wet mill product purchased from a local market. Tamarind gum was obtained from a manufacturing company (Socius Ingredients LLC, Evanston, IL, USA), and sugar (CJ CheilJedang Corp, Seoul, Korea), soy milk powder (Benesoy, Rock City, IL, USA), canola oil (Beksul, Seoul, Korea), salt (Haepyo, Seoul, Korea), and dry yeast (Lesaffre, Marcqen–Baroeul, France) were purchased from a local market. The anti-staling enzyme Novamyl 10,000 BG, used in the storage experiments, was obtained from a manufacturing company (Novozymes, Bagsvaerd, Denmark).

2.2. Preparation of Rice Bread

The Design Expert (Minneapolis, MN, USA) program was used for the factorial design to identify and optimize the significant factors of the formulation and processing conditions. The four factors—concentration of gum, amount of water, mixing time, and fermentation time—were selected, with each factor consisting of two levels: gum concentration at 1 and 2%; water at 80 and 100 g; mixing time at 5 and 10 min; and fermentation time at 30 and 60 min. The basic ingredients and formulation used were 100 g of rice flour, 5 g of canola oil, 7 g of sugar, 3 g of soy milk powder, 2 g of salt, and 3 g of yeast, according to the combined method described by Hera et al. [16] and Jang et al. [14]. The rice bread was prepared as per the method described by Jang [17].

Pre-weighed dry ingredients, including rice flour and tamarind gum, were tumbled in a jar with a lid for uniform blending and transferred to a mixer bowl (Pin Mixer, National Mfg. Co., Lincoln, NE, USA). Oil (5 g) and water (amount set as per the experimental design shown in Table 1) were added and mixed for a set time. Batter (190 g) was placed in a baking pan (a non-stick mini loaf pan, 14.6 cm × 8.3 cm × 5.7 cm; Chicago Metallic, Lake Zurich, IL, USA). The baking pan containing the batter was placed in a fermenter (Samjung, Gyeonggi, Korea) for a set time at 30 °C and 85% relative humidity. Batter appearance, before and after fermentation, was photographed using a digital camera (Canon, Tokyo, Japan). After fermentation, 10 g of the batter was removed for pH measurement, and the remaining batter was baked in an oven (Phantom M301 Combi, Samjung, Gyeonggi, Korea) at 215 °C for 20 min. After cooling at room temperature (25–30 °C) for 20 min, the bread was taken out of the baking tray. It was further cooled for 40 min before quality analysis. Bread baking was repeated twice for each condition.

Table 1. Experimental conditions tested by a full factorial design for preparing gluten-free rice bread.

Std Order	Run Order	Gum (%)	Water (g)	Mixing Time (min)	Fermentation Time (min)
1	12	1	80	5	30
2	14	2	80	5	30
3	9	1	100	5	30
4	16	2	100	5	30
5	3	1	80	10	30
6	5	2	80	10	30
7	10	1	100	10	30
8	17	2	100	10	30
9	13	1	80	5	60
10	7	2	80	5	60
11	11	1	100	5	60
12	2	2	100	5	60
13	1	1	80	10	60
14	6	2	80	10	60
15	4	1	100	10	60
16	15	2	100	10	60
17	8	1.5	90	7.5	45

2.3. Analysis of Batter Property of Rice Bread

To measure the specific gravity of the batter, 5 g of the batter was placed in a 50 mL beaker, and its volume was marked. After removing the batter, water was poured up to the mark, and the volume and weight of the water were measured. The specific gravity of the batter was calculated.

To measure the pH before and after fermentation of the batter, 5 g of batter and 50 g of distilled water were mixed in a blender (SFM-7700JJH, Sinil, Korea) for 20 s to ensure uniform dispersion, and the batter suspension was transferred to a beaker. The pH of the suspension was measured using a pH meter (Mettler Toledo SevenEasy pH meter S20, Columbus, OH, USA).

2.4. Evaluation of Rice Bread Quality

Bread weight, height, and volume were measured according to the method described by Jang et al. [14]. To measure bread volume, millet seeds were used in the seed replacement method. The density (g/mL) of millet, used for converting weight into volume, was determined by filling a 1.5 L plastic container with millet. A loaf of bread was placed in the container filled with millet, and the weight of millet that was pushed out of the container (g) was measured. The volume of the bread (mL) was calculated based on the calculated density.

According to the AACC Approved Method 74–09.01 [18] and that described by Jang et al. [14], bread firmness was measured using a Texture Analyzer (Brookfield CT3, Middleboro, MA, USA). The measurement conditions were set as follows: mode, measure force in compression; pre-test speed, 2.0 mm/s; test speed, 2.0 mm/s; post-test speed, 5.0 mm/s; probe, a 3.6 cm diameter cylinder; and penetration distance, 15 mm. One loaf of bread was cut from the centre into two pieces with a thickness of 2 cm; its firmness was measured by compressing the centre of the cross-section of the bread.

The moisture content of the bread samples was measured using the AACC Approved Method 44–15.02 [18]. Approximately 3 g of pre-weighed breadcrumbs in an aluminium container was placed in a hot-air drying oven (FO 600-M, Jeio Tech, Daejeon, Korea), dried at 130 °C for 1 h, and cooled at room temperature in a desiccator for 30 min. The moisture content of the bread was calculated from the weight reduced after drying.

2.5. Evaluation of the Effect of an Anti-staling Enzyme on the Staling of Rice Bread during Storage

To evaluate the effect of an anti-staling enzyme on bread staling during storage, bread was prepared as per the formulation and processing conditions were selected as the most desirable in the optimization experiment conducted in this study. An anti-staling enzyme, Novamyl 10,000 BG, was added to the bread formulation at 0.005, 0.01, and 0.015%, and mixed well with the dry ingredients used to prepare the rice bread as per the abovementioned process. Control bread, for comparison, was made without adding the enzyme. One loaf of bread was cut from the centre into three pieces with a thickness of 2 cm. One piece was used as the fresh sample (0 days); the other two pieces were sealed in aluminium foil packaging to avoid moisture loss during storage and placed in a refrigerator at 4 °C. Measurements of firmness, moisture content, and enthalpy of starch retrogradation of bread by DSC during storage were performed on 0, 1, and 4 days of storage after baking.

The temperature and enthalpy of starch retrogradation of bread were measured by DSC (DSC 6000, Perkin–Elmer Co., Waltham, MA, USA). Breadcrumbs and water were mixed well at a ratio of 1:1 (*w*/*w*). Then, 40 mg of the mixture was sealed in a stainless steel pan and heated at 10 °C/min of heating rate from 10 °C to 140 °C using a differential scanning calorimeter. The temperature and enthalpy of starch retrogradation of bread were analyzed using Pyris Software (ver. 11, Perkin–Elmer Co.).

2.6. Statistical Analysis

All results were evaluated by two or more repeated experiments and analyzed using the Design Expert program to identify significant factors, as well as to optimize the formulation and processing conditions. The SPSS statistical program (ver. 22.0, IBM Corp., Armonk, NY, USA) was used for the ANOVA and the Duncan's multiple range test at a significance threshold of $p < 0.05$.

3. Results and Discussion

3.1. Characteristics of the Batter of Rice Bread

The appearance of the batter, prepared according to the experimental design, is presented in Figure 1. With the high WA in the formulation, the batter was thin. Within the same WA in the formulation, the batter with the high GC was relatively thicker. With the increasing WA and FT, the expanded volume of the batter after fermentation increased. The most significant increase in batter volume after fermentation was observed in batters #11, 12, 15, and 16, which were formulated with 100 g of water and fermented for 60 min, and #17, the center point with 90 g of water for 45 min of fermentation. Increasing the FT could break the dough structure by weakening the hydrocolloid network in which starch granules or flour particles are embedded, resulting in less resistance to strain [19]. The extent of volume expansion between these batter samples was not significantly different by varying the MT. However, when the GC was less, the batter looked thin, which might have resulted in easy expansion. A lower batter consistency is preferred for good performance during leavening [20]. Thus, the appearance of the batter might be linked to the bread volume.

The pH of the rice bread batter, before and after fermentation, is shown in Table 2. After fermentation, the pH of all the samples decreased, which was considered to be the result of acidic substances, such as lactic acid and acetic acid, that were produced during fermentation [21]. The pH change during fermentation was 0.3–0.7. The batter samples (#1–8), fermented for 30 min, showed a pH of 5.7–5.9 after the fermentation, resulting in a pH change of 0.3–0.5, whereas the samples (#9–16), fermented for 60 min, showed a pH of 5.4–5.6 after fermentation, resulting in a pH change of 0.5–0.7. As a result, it is suggested that the fermentation time affected the batter pH, whereas the GC, WA, and MT did not show any significant effects on batter pH. Jang et al. [14] reported that changes in batter pH during fermentation were different between various gums used at 4%. However, no significant difference was observed at the 1–2% concentration of tamarind gum that was used in this study in order to follow the U.S. Food and Drug Administration (FDA) regulation.

Figure 1. Appearance of batter prepared as per the designed formula and processing conditions before (**left**) and after (**right**) fermentation.

The specific gravity of all of the batter samples was 1.1 g/cc, which indicated no significant difference in aeration during mixing due to a negligible difference (1 and 2%) in the viscosity of the gum. But, Jang [17] observed differences in the specific gravity of the batter depending on the type of gum used in the gluten-free rice bread. When the hydroxypropyl methylcellulose (HPMC) concentration was increased from 1 to 4% in the formulation of rice bread, the specific viscosity of the batter was significantly increased because of the increased thickness that caused less aeration during mixing.

3.2. Quality Characteristics of Rice Bread

The cross-section of the rice bread is shown in Figure 2. The #9–16 bread samples with 60 min of fermentation exhibited a relatively higher volume than the #1–8 bread samples with 30 min of fermentation, reflecting the effect of FT on the bread volume. The impact of the WA was more pronounced at a longer FT, and the #11, 12, 15, and 16 bread samples exhibited relatively large cross-sectional volumes of bread. Among the bread samples, #11 (1% gum, 100 g of water, 5 min MT, and 60 min FT) was selected as the most desirable bread that was similar to commercial bread, based on crumb pore size and symmetrical shape.

The weight of the bread was affected significantly by the high WA with 60 min FT but not with 30 min FT. The weight of the former bread per loaf was 144.1–145.6 g, whereas that of the latter was 150.3–153.2 g (Table 2). When the WA was high and the FT long, the batter could expand quickly during the baking process, resulting in an increase in volume and moisture loss during baking [5, 16]. The volume of rice bread was also affected significantly by the FT with high WA. At 60 min of fermentation, the volume of bread was 338.3–380.0 mL, whereas, at 30 min, the volume of bread was 270.6–293.5 mL (Table 2). When the WA was low, the effect of the FT on bread volume was not significant.

Figure 2. Cross-sections of gluten-free rice bread tested by a full factorial design.

The moisture content and firmness of the breadcrumbs are shown in Table 2. The moisture content of the breadcrumbs correlated with the amount of water in the formulation. Regardless of the FT, the water content of the crumbs was 44.4–46.3% when a small amount of water was added to the formulation and 49.3–51.3% when a large amount of water was added to the formulation. The firmness of the breadcrumbs showed a significant relationship with the moisture content of the breadcrumbs in the bread samples prepared with the same FT. When the FT is long (60 min) and the WA is large, the bread volume is large and the firmness of the bread crumb is significantly low [5,16,22]. Overall, the quality of bread was affected by the WA and FT, which were the significant factors, but was not affected by the GC and MT.

Table 2. pH of batter before and after fermentation and gluten-free rice bread characteristics tested by a full factorial design.

Std Order	Batter		Bread			
	pH before Fermentation	pH after Fermentation	Weight (g)	Volume (mL)	Moisture Content (%)	Firmness (N)
1	6.2 ± 0.1 [a,*]	5.9 ± 0.1 [a]	153.8 ± 0.2 [b,c]	289.8 ± 11.5 [b]	46.2 ± 0.0 [a]	9.8 ± 1.1 [e,f]
2	6.3 ± 0.2 [a]	5.9 ± 0.2 [a]	154.1 ± 0.5 [b,c]	300.1 ± 0.9 [b]	45.3 ± 0.3 [a]	8.9 ± 0.1 [d,e,f]
3	6.2 ± 0.0 [a]	5.9 ± 0.1 [a]	150.7 ± 0.2 [b,c]	273.2 ± 5.0 [a,b]	50.0 ± 0.6 [c]	7.0 ± 1.2 [b,c,d,e]
4	6.3 ± 0.2 [a]	5.9 ± 0.2 [a]	153.2 ± 0.7 [b,c]	370.1 ± 6.0 [d,e]	49.9 ± 0.5 [c]	6.8 ± 1.2 [b,c,d,e]
5	6.0 ± 0.1 [a]	5.7 ± 0.1 [a]	152.9 ± 0.9 [b,c]	295.2 ± 5.0 [b]	45.6 ± 0.7 [a]	8.5 ± 0.9 [c,d,e]
6	6.0 ± 0.1 [a]	5.7 ± 0.2 [a]	154.6 ± 1.1 [c]	276.2 ± 14.1 [a,b]	44.4 ± 0.3 [a]	8.8 ± 1.0 [d,e,f]
7	6.2 ± 0.0 [a]	5.8 ± 0.0 [a]	152.3 ± 0.8 [b,c]	258.3 ± 4.5 [a]	51.3 ± 0.1 [c]	7.0 ± 0.6 [b,c,d,e]
8	6.3 ± 0.3 [a]	5.8 ± 0.2 [a]	152.2 ± 1.4 [b,c]	274.3 ± 8.8 [a,b]	50.6 ± 0.8 [c]	7.0 ± 0.2 [b,c,d,e]
9	6.2 ± 0.0 [a]	5.5 ± 0.0 [a]	151.1 ± 0.3 [b,c]	282.7 ± 3.5 [a,b]	46.3 ± 0.3 [a]	7.9 ± 1.0 [c,d,e]
10	6.1 ± 0.2 [a]	5.5 ± 0.2 [a]	152.0 ± 0.1 [b,c]	270.6 ± 6.2 [a,b]	44.4 ± 0.2 [a]	9.9 ± 1.1 [e,f]
11	6.2 ± 0.1 [a]	5.6 ± 0.0 [a]	145.3 ± 1.1 [a]	338.3 ± 3.8 [c]	50.6 ± 0.7 [c]	4.0 ± 0.1 [a,b]
12	6.1 ± 0.2 [a]	5.5 ± 0.1 [a]	145.2 ± 0.6 [a]	360.0 ± 12.4 [c,d,e]	50.4 ± 1.0 [c]	5.5 ± 0.4 [a,b,c]
13	6.0 ± 0.1 [a]	5.5 ± 0.1 [a]	153.2 ± 4.1 [b,c]	293.5 ± 12.8 [b]	45.2 ± 1.3 [a]	12.0 ± 2.0 [f]
14	6.0 ± 0.1 [a]	5.4 ± 0.2 [a]	150.3 ± 0.6 [b]	275.3 ± 0.8 [a,b]	44.4 ± 0.2 [a]	8.5 ± 0.9 [c,d,e]
15	6.1 ± 0.1 [a]	5.4 ± 0.2 [a]	144.1 ± 0.2 [a]	349.1 ± 8.0 [c,d]	49.3 ± 0.3 [b,c]	3.1 ± 0.3 [a]
16	6.3 ± 0.3 [a]	5.6 ± 0.2 [a]	145.6 ± 1.0 [a]	380.0 ± 10.9 [e]	50.6 ± 0.3 [c]	4.0 ± 1.2 [a,b]
17	6.0 ± 0.2 [a]	5.5 ± 0.2 [a]	149.9 ± 0.5 [b]	365.6 ± 18.8 [c,d,e]	47.0 ± 2.4 [a,b]	5.8 ± 1.1 [a,b,c,d]

* Results are expressed as the mean ± standard deviation. Values with the same letter within the same column are not significantly different ($p < 0.05$), according to Duncan's multiple range test.

3.3. Significant Factors Affecting the Quality Characteristics of Rice Bread

The results of the statistical analysis of the batter characteristics and bread quality characteristics of gluten-free rice bread prepared based on the full factorial design of a factor analysis method are shown in Table 3. The three-dimensional plots of the six characteristics of gluten-free rice batter and bread with tamarind gum are presented as a function of the two independent variables among GC, WA, MT, and FT in response to their significance in Figure 3.

Table 3. Significant factors for the batter and bread characteristics of gluten-free rice bread tested by a full factorial design.

Sample	Response	Significant Factor	*p*-Value	Significance
Batter	Specific gravity	-	-	-
	pH before fermentation	-	-	-
	pH after fermentation	Fermentation time (FT)	0.0002	Negative
Bread	Weight	Fermentation time	<0.0001	Negative
		Water amount (WA)	<0.0001	Positive
		WA × FT	0.0024	Positive
	Volume	Fermentation time	0.0207	Positive
		WA × FT	0.0097	Positive
	Moisture content	Water amount	0.0002	Positive
	Firmness	Water amount	<0.0001	Negative

With regard to the batter characteristics, no significant factors affected the specific gravity (Figure 3A) or pH before fermentation; however, FT was a significant factor that was negatively correlated with the pH after fermentation. It was predicted that the pH of the batter decreased after fermentation with the increase in the FT (Figure 3B).

With regard to the quality characteristics of bread, the WA and FT were the significant factors negatively correlated with the weight of bread. The weight of the bread decreased as the WA and FT increased (Figure 3C). However, FT was positively correlated with bread volume, confirming that the FT needs to be controlled to increase the volume. Interaction of WA and FT also had a positive effect on bread volume. Thus, the bread volume increased as the WA and FT increased (Figure 3D). The moisture content of breadcrumbs was positively correlated with the WA, whereas the firmness of the breadcrumbs was negatively correlated with the WA (Figure 3E,F). These results reflected the importance of the WA for bread quality. Hera et al. [16] and Mancebo et al. [5] reported the effect of water content on gluten-free bread quality; the bread specific volume increased and the crumb texture improved by increasing dough hydration, which was similar to our results.

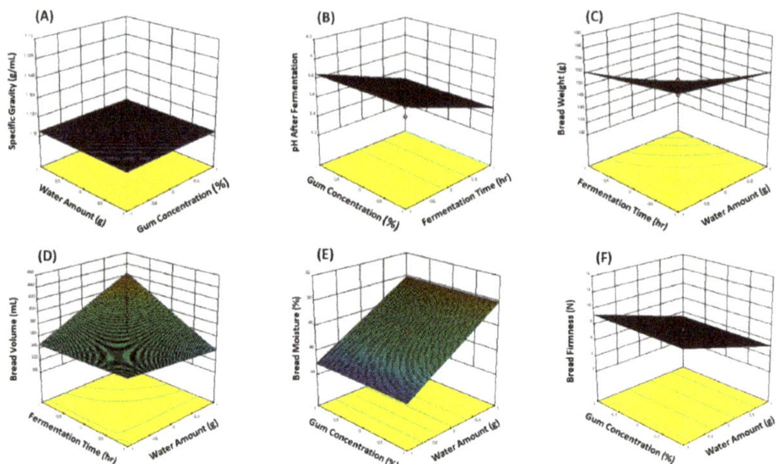

Figure 3. Three-dimensional plots on the characteristics of batter (**A,B**) and bread (**C–F**) affected by significant factors of the formula and processing conditions in the preparation of gluten-free rice bread with tamarind gum.

3.4. Effect of an Anti-staling Enzyme on the Quality of Rice Bread during Storage

The moisture content and firmness of bread after adding an anti-staling enzyme during storage at 4 °C and its retrogradation tendency measured by DSC are shown in Figure 4 and Table 4. The control rice bread without the added enzyme showed a significant increase in firmness during storage compared to the bread with the added enzyme. The results of the control bread were consistent with the storage test of rice bread with tamarind gum reported by Jang et al. [14]. Gluten-free rice bread has a higher starch percentage than wheat bread, and numerous attempts to retard staling have been reported [23,24].

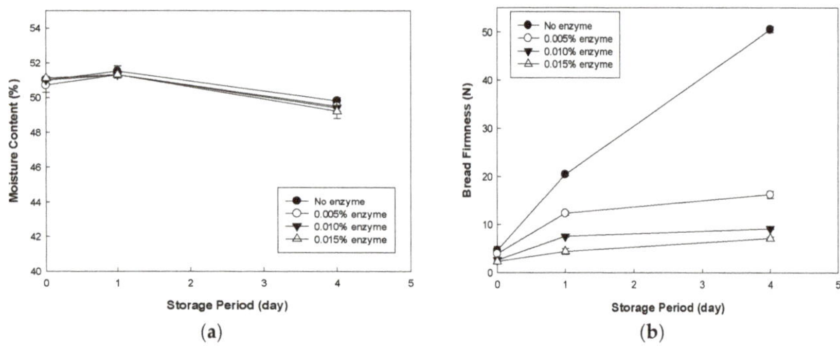

Figure 4. Changes in the moisture content (**a**) and firmness (**b**) of bread with anti-staling enzyme at different concentrations during storage at 4 °C.

Table 4. Thermal characteristics of gluten-free rice bread prepared with #11 (1% gum, 100 g of water, a 5 min mixing time, and a 60 min fermentation time) during storage at 4 °C.

Enzyme Concentration (%)	Storage Time (day)	Tonset (°C)	Tpeak (°C)	Tend (°C)	Heat of Transition (ΔQ, J/g)
0	0	40.1 ± 0.1 [a,*]	48.1 ± 0.1 [a]	66.9 ± 1.1 [c,d,e,f]	0.35 ± 0.0 [a]
	1	40.0 ± 0.0 [a]	53.4 ± 0.1 [b]	64.8 ± 0.2 [b,c,d,e]	0.82 ± 0.0 [c]
	4	40.8 ± 0.9 [a]	52.8 ± 0.4 [b]	65.1 ± 0.6 [b,c,d,e]	1.43 ± 0.1 [e]
0.005	0	40.0 ± 0.0 [a]	48.0 ± 0.1 [a]	67.8 ± 0.9 [e,f]	0.28 ± 0.0 [a]
	1	40.0 ± 0.0 [a]	52.8 ± 0.6 [b]	64.6 ± 0.0 [b,c,d]	0.60 ± 0.0 [b]
	4	40.7 ± 0.3 [a]	52.8 ± 0.6 [b]	64.6 ± 0.7 [b,c,d]	1.11 ± 0.1 [d]
0.010	0	40.18 ± 0.1 [a]	48.7 ± 0.1 [a]	68.4 ± 0.3 [f]	0.37 ± 0.0 [a]
	1	40.0 ± 0.0 [a]	52.7 ± 0.2 [b]	61.0 ± 0.8 [a]	0.40 ± 0.1 [a]
	4	40.47 ± 0.1 [a]	51.4 ± 2.1 [b]	64.6 ± 0.6 [b,c,d]	1.02 ± 0.1 [d]
0.015	0	40.0 ± 0.0 [a]	48.5 ± 1.2 [a]	67.6 ± 2.4 [d,e,f]	0.25 ± 0.0 [a]
	1	40.0 ± 0.0 [a]	53.0 ± 0.5 [b]	63.8 ± 0.7 [a,b,c]	0.36 ± 0.1 [a]
	4	40.79 ± 0.7 [a]	52.7 ± 0.8 [b]	63.5 ± 0.0 [a,b]	0.73 ± 0.0 [b,c]

* Results are expressed as the mean ± standard deviation. Values with the same letter within the same column are not significantly different ($p < 0.05$), according to Duncan's multiple range test.

The firmness of the control rice bread before refrigerated storage was 4.6 N and that of the bread with the added enzyme was slightly less (2.3–3.9 N) (Figure 4). With the increasing amount of enzyme, the firmness of the bread decreased. With the increasing storage period, the firmness increased. The difference between the control and the enzyme addition group was more significant on day four. The firmness of bread after refrigerated storage increased from 4.6 to 50.5 N for the control group, from 3.9 to 16.2 N for the 0.005% enzyme group, from 2.6 to 9.1 N for the 0.010% enzyme group, and from 2.3 to 9.1 N for the 0.015% enzyme group. As the amount of enzyme increased, the extent of the increase in firmness during storage significantly decreased ($p < 0.05$), demonstrating the effective anti-staling impact of the enzyme. The reaction of maltogenic amylase is known to hydrolyze starch in the batter during mixing and baking and reduce the size of amylose and amylopectin. Shorter amylose and amylopectin cannot rearrange easily, resulting in retardation of starch retrogradation [25].

The moisture content of the rice bread during refrigerated storage was 51.1–49.2% (Figure 4), which was the most favorable condition for starch retrogradation [26]. During storage, it showed a slight change on the fourth day for all samples, indicating the excellent packaging of the bread during storage. The bread with the enzyme showed less of an increase in firmness than the control bread, demonstrating the anti-staling effect of the enzyme during storage.

The enthalpy of starch retrogradation of bread with and without the enzyme during storage, measured by DSC, was compared (Table 4). As the storage period increased, the peak of the control bread in the DSC thermogram increased significantly; however, that of the bread with the enzyme did not increase dramatically. Moreover, as the amount of enzyme increased, the increase in peak size during storage decreased. The enthalpy of starch retrogradation of bread increased with an increase in the storage period. The control bread showed the highest increase in enthalpy, whereas that with the enzyme showed a less significant increase in enthalpy. The extent of the increase in enthalpy decreased as the amount of added enzyme increased. The retrogradation enthalpy on the fourth day of storage was 1.43, 1.11, 1.02, and 0.73 J/g for the control bread and 0.005, 0.010, and 0.015% enzyme added bread, respectively. α-Amylase hydrolyzes starch while increasing sugars, which can be fermented by the yeast. The enzyme reaction increases the bread volume, improves the color and flavor of the bread crust, and extends the shelf life by retarding staling [27]. In this study, the addition of the enzyme inhibited the increase in firmness of gluten-free rice bread during storage, indicating the excellent anti-staling effect. This finding suggests its potential for successful application in commercially manufactured rice bread.

4. Conclusions

A factorial analysis was used to identify and optimize significant factors in the formulation and processing conditions of rice bread with tamarind gum. Four factors—gum concentration, water amount, mixing time, and fermentation time—were selected and two levels for each factor were set, making a total of 17 conditions, including the center point. Although WA and FT significantly affected the quality of gluten-free rice bread, GC and MT did not. Overall, the WA and FT were identified as the significant factors in the production of gluten-free rice bread. Therefore, proper control over WA and FT can maximize the bread volume and minimize the firmness of the bread. Among the various experimental conditions, 1% tamarind gum, 100 g of water, 5 min of mixing, and 60 min of fermentation were determined to be the optimal conditions. The addition of the anti-staling enzyme was effective in retarding the increase in retrogradation enthalpy and bread firmness. Using an optimized formula and processing factors for gluten-free rice bread with the combined addition of tamarind gum and an anti-staling enzyme can be applied successfully in commercially manufactured gluten-free rice bread.

Author Contributions: Conceptualization, Y.-E.H. and M.K.; Data curation, Y.-E.H.; Data analysis, Y.-E.H. and M.K.; Writing—original draft, Y.-E.H.; Writing—review & editing, M.K. All authors have read and agreed to the published version of the manuscript.

Acknowledgments: We thank Stevan Angalat at Angalat Group International LLC for his kind support in requesting the soy milk powder and the tamarind gum from the manufacturers for the study.

Conflicts of Interest: The authors declare no conflicts of interest.

References

1. Statistic Korea. Annual Trends of Food Grain Consumption in 2018. Available online: http://kostat.go.kr/portal/eng/pressReleases/1/index.board?bmode=read&aSeq=373404 (accessed on 1 December 2019).
2. Green, P.H.R.; Jabri, B. Coeliac disease. *Lancet* **2003**, *362*, 383–391. [CrossRef]
3. Kim, J.M. Characteristics of Rice Flours Prepared from Different Rice Varieties and Development of Gluten-Free Bakery Products Using Them. Ph.D. Thesis, Chonnam National University, Gwangju, Korea, February 2016.
4. Han, H.M.; Cho, J.H.; Kang, H.W.; Koh, B.K. Rice varieties in relation to rice bread quality. *J. Sci. Food Agric.* **2012**, *92*, 1462–1467. [CrossRef]
5. Mancebo, C.M.; Miguel, M.A.S.; Martinez, M.M.; Gomez, M. Optimisation of rheological properties of gluten-free doughs with HPMC, psyllium and different levels of water. *J. Cereal Sci.* **2015**, *61*, 8–15. [CrossRef]
6. Mohammadi, M.; Sadeghnia, N.; Azizi, M.H.; Neyestani, T.R.; Mortazavian, A.M. Development of guten-free flat bread using hydrocolloids: Xanthan and CMC. *J. Ind. Eng. Chem.* **2014**, *20*, 1812–1818. [CrossRef]
7. Masure, H.G.; Fierens, E.; Delcour, J.A. Current and forward looking experimental approaches in gluten-free bread making research. *J. Cereal Sci.* **2016**, *67*, 92–111. [CrossRef]
8. Witczak, M.; Ziobro, R.; Juszczak, L.; Korus, J. Starch and starch derivatives in gluten-free systems—A review. *J. Cereal Sci.* **2016**, *67*, 46–57. [CrossRef]
9. Anton, A.A.; Artifield, S.D. Hydrocolloids in gluten-free breads: A review. *Inter. J. Food Sci. Nutr.* **2008**, *59*, 11–23. [CrossRef] [PubMed]
10. Lazaridou, A.; Duta, D.; Papageorgiou, M.; Belc, N.; Biliaderis, C.G. Effects of hydrocolloids on dough rheology and bread quality parameters in gluten-free formulations. *J. Food Eng.* **2007**, *79*, 1033–1047. [CrossRef]
11. Mir, S.A.; Shah, M.A.; Naik, H.R.; Zargar, I.A. Influence of hydrocolloids on dough handling and technological properties of gluten-free breads. *Trends Food Sci. Tech.* **2016**, *51*, 49–57. [CrossRef]
12. Nishita, K.D.; Roberts, R.L.; Bean, M.M. Development of a yeast-leavened rice-bread formula. *Cereal Chem.* **1976**, *53*, 626–635.
13. Sciarini, L.S.; Ribotta, P.D.; León, A.E.; Pérez, G.T. Effect of hydrocolloids on gluten- free batter properties and bread quality. *Int. J. Food Sci. Tech.* **2010**, *45*, 2306–2312. [CrossRef]
14. Jang, K.J.; Hong, Y.E.; Moon, Y.J.; Jeon, S.J.; Angalat, S.; Kweon, M. Exploring the applicability of tamarind gum for making gluten-free rice bread. *Food Sci. Biotechnol.* **2018**, *27*, 1639–1648. [CrossRef] [PubMed]

15. Maeda, T.; Yamashita, H.; Morita, N. Application of xyloglucan to improve the gluten membrane on breadmaking. *Carbohydr. Polym.* **2007**, *68*, 658–664. [CrossRef]

16. Hera, E.; Martinez, M.; Gomez, M. Influence of flour particle size on quality of gluten-free rice bread. *LWT—Food Sci. Technol.* **2013**, *54*, 199–206. [CrossRef]

17. Jang, K.J. Effect of Tamarind Gum on Gluten-Free Rice Batter Properties and Bread Quality. Master's Thesis, Pusan National University, Busan, Korea, February 2018.

18. AACC. Method 44-15.02, Method 74-09.01.2010. In *Approved Methods of Analysis*, 11th ed.; American Association of Cereal Chemists: St. Paul, MN, USA.

19. Martinez, M.M.; Gomez, M. Rheological and microstructural evolution of the most common gluten-free flours and starches during bread fermentation and baking. *J. Food Eng.* **2017**, *197*, 78–86. [CrossRef]

20. Cappa, C.; Lucisano, M.; Mariotti, M. Influence of Psyllium, sugar beet fibre and water on gluten-free dough properties and bread quality. *Carbohydr. Polym.* **2013**, *98*, 1657–1666. [CrossRef]

21. Blanco, C.A.; Ronda, F.; Pérez, B.; Pando, V. Improving gluten-free bread quality by enrichment with acidic food additives. *Food Chem.* **2011**, *127*, 1204–1209. [CrossRef]

22. Tsatsaragkou, K.; Gounaropoulos, G.; Mandala, I. Development of gluten free bread containing carob flour and resistant starch. *LWT—Food Sci. Technol.* **2014**, *58*, 124–129. [CrossRef]

23. Bollain, C.; Collar, C. Dough viscoelastic response of hydrocolloid/enzyme/surfactant blends assessed by uni- and bi-axial extension measurements. *Food Hydrocoll.* **2004**, *18*, 499–507. [CrossRef]

24. Gujral, H.S.; Rosell, C.M. Improvement of the breadmaking quality of rice flour by glucose oxidase. *Food Res. Int.* **2004**, *37*, 75–81. [CrossRef]

25. Ribotta, P.D.; Perez, G.T.; Leon, A.E.; Anon, M.C. Effect of emulsifier and guar gum on micro structural, rheological and baking performance on frozen bread dough. *Food Hydrocoll.* **2004**, *18*, 305–313. [CrossRef]

26. Zeleznak, K.J.; Hoseney, R.C. The role of water in the retrogradation of wheat starch gels and bread crumb. *Cereal Chem.* **1986**, *63*, 407–411.

27. Collar, C.; Bollain, C.; Angioloni, A. Significance of microbial transglutaminase on the sensory, mechanical and crumb grain pattern of enzyme supplemented fresh pan breads. *J. Food Eng.* **2005**, *70*, 479–488. [CrossRef]

Article

Physical and Dynamic Oscillatory Shear Properties of Gluten-Free Red Kidney Bean Batter and Cupcakes Affected by Rice Flour Addition

Pavalee Chompoorat [1,*], Napong Kantanet [1], Zorba J. Hernández Estrada [2,3] and Patricia Rayas-Duarte [2]

[1] Faculty of Engineering and Agro-Industry, Maejo University, Chiang Mai 50290, Thailand; napong250142@gmail.com

[2] Robert M Kerr Food & Agricultural Products Center, Department of Biochemistry and Molecular Biology, Oklahoma State University, Stillwater, OK 74078, USA; zorba.hernandez@itver.edu.mx (Z.J.H.E.); pat.rayas_duarte@okstate.edu (P.R.-D.)

[3] Tecnológico Nacional de México/I.T. Veracruz, Calz. Miguel Angel de Quevedo No. 2779 Col. Formando Hogar, Veracruz 91860, Mexico

* Correspondence: pavalee@mju.ac.th; Tel.: +669-3576-3571

Received: 31 March 2020; Accepted: 7 May 2020; Published: 11 May 2020

Abstract: Red kidney bean (RKB) flour is a nutrient-rich ingredient with potential use in bakery products. The objective of this study was to investigate the viscoelastic properties and key quality parameters of a functional RKB flour in gluten-free cupcakes with different rice flour levels. A 10 g model batter was developed for analyzing the viscoelastic properties of RKB with rice incorporation, in a formula containing oil, liquid eggs, and water. Rice flour was added at five levels 0%, 5%, 10%, 15%, and 25% (*w/w*, g rice flour/100 g RKB flour). Rice flour increased RKB batter consistency, solid- and liquid-like viscoelastic behavior and revealed a heterogeneous structure, based on the sweep frequency test. Rice flour at the 25% level increased the shear modulus and activation energy of gelatinization, compared to 0% rice flour addition. Rice flour levels in the RKB batter decreased the inflection gelation temperature from 63 to 56 °C. In addition, the texture of RKB cupcakes with 25% rice flour were 46% softer, compared to the control. The scores from all sensory attributes of cupcakes increased with the addition of rice flour. Rice flour addition improved solid- and liquid-like behavior of the RKB batter and improved the cupcake's macro-structural characteristics. Overall, 25% rice flour addition performed better than the lower levels. This study confirmed the potential of RKB as a functional ingredient and its improvement in cupcake application with the addition of rice flour.

Keywords: dynamic oscillatory shear test; non-isothermal kinetic modeling; gluten-free cupcake; red kidney bean

---

1. Introduction

Red kidney bean (*Phaseolus vulgaris L.*) is a good source of vegetable protein, insoluble and soluble fiber, and is overall a nutrient-rich food ingredient with a great potential for different food applications [1]. Thermal properties, water and oil retention as well as the emulsifying properties of flour of red kidney bean have been reported. Processing including vacuuming and soaking decrease the beany flavor profile of the RKB flour. [2]. However, more studies are needed to fully understand the rheological and sensory properties of the red kidney bean flour and its potential use in bakery products.

Commonly, gluten-free products are low in nutritional value [3,4], with high carbohydrate, low fiber, high glycemic index, as well as saturated and hydrogenated fatty acids [4]. An example of alternative flours and ingredients that increase the protein and bioactive compounds is germinated brown rice flour; a 48 h germination time increases the γ-aminobutyric acid (GABA), polyphenols levels, and antioxidant activity [5]. Unhusked buckwheat flour enhances the fiber content and texture properties of gluten-free bread [6]. However, the use of this ingredient is limited due to its high insoluble fiber from husk, which affects the sensory properties of bread [6]. Rice and tapioca flours are widely used as the main ingredients in gluten-free bakery products due to their neutral color and flavor [7,8].

There is an increased interest in using pulses in food ingredients, including FAO declaring 2016 as the International Year of Pulses with the goal of encouraging researchers around the world to investigate more applications of pulses. Pea and chickpea flours have been proposed as alternative flour ingredients in the formulation of gluten-free products [9]. Protein isolates from kidney bean, field pea, and amaranth incorporated in rice-based gluten-free muffins increased the batter viscoelasticity and improved the specific volume and texture [9]. Two types of cowpea protein isolate used in rice flour muffins affected the gluten-free muffin volume differently [10]. The incorporation of the white cowpea protein isolate improved the volume of gluten-free muffin while red cowpea protein isolate decreased the volume [10]. Studies on a mechanistic model to provide a better understanding of baking was addressed by other researchers, including heat and mass transfer during baking and a model for baking of leavened aerated food [11–13]. It was proposed that bubble growth during heating showed a lag time before the exponential growth phase, which is related to an expanded inner region of more or less uniform density. The study of rheometric non-isothermal gelatinization kinetics also proposed to model an alteration of gluten-free muffin batter and found that zero-order reaction kinetics can be used to explain batter gelatinization [14]. To the best of our knowledge, the use of the whole red kidney bean flour in gluten-free cupcakes with rice flour has not been investigated. Therefore, the aim of this work was to examine the effect of rice flour on gluten-free RKB batter rheological properties and cupcake quality properties, as well as the interrogation of possible correlations between these properties.

2. Materials and Methods

2.1. Flour Preparation

Red kidney beans and rice flour were purchased in a local market in Chiang Mai, Thailand. Red kidney beans (RKB) were rinsed, boiled in water for 20 min, strained, and dried for 4 h at 80 °C using a convection oven. These conditions were selected from Chompoorat et al. (2018) [15]. The resulting beans were ground into flour with a hammer mill (W.J. Fit Company, Chicago, IL, USA) to a particle size less than 250 μm (stainless steel sieve # 60, W.S. Tyler Co., Mentor, OH, USA). The RKB flour was stored in polyethylene bags at 4 °C, until needed. RKB flour had 20.5% protein and 14.7% fiber, as previously reported [15]; while, rice flour had 80% carbohydrate, 3.5% protein, and 0.89% fat, according to USDA [16].

2.2. Cupcake Preparation

Gluten-free RKB cupcakes were prepared with ingredients purchased from a local market and RKB flour produced in the laboratory. The cupcake formula expressed as a weight percent of RKB flour base (100%) contained granulated white sugar (65%), baking powder (1.9%), baking soda (1.1%), cocoa powder (10%), salt (1%), guar gum (0.5%), vegetable oil (37%), egg (65%), water (150%), and vanilla extract (3%). Commercial rice flour was added at 0%, 5%, 10%, 15%, and 25% based on RKB flour. The control sample contained 100% RKB flour.

The cupcakes were prepared by mixing all dry ingredients (KitchenAid, St. Joseph, MI, USA) for 1 min at the lowest speed (setting# 1), using a paddle attachment. Wet ingredients (vegetable oil, egg, water, and vanilla) were added, mixed for 2 min at low speed (setting #1), followed by 5 min at high speed (setting # 6). Cupcake batter aliquots (40 g) were placed into paper-lined molds and baked in a convection oven at 195 °C for 20 min. The cupcakes were removed from the mold, allowed to cool for

60 min at room temperature, and immediately transferred for analysis of firmness inside a cupcake cardboard box.

2.3. Batter Consistency

Batter aliquot (100 g) was used for consistency measurements with a Bostwick consistometer (CSC Scientific Company, Inc., Fairfax, VA, USA), following manufacturer's directions. The cupcake batter was transferred to the consistometer chamber, the gate was released, and the batter flow rate (cm per min) was recorded. Batter flow rate was inversely proportional to consistency. Measurements were made in three replicates per treatment.

2.4. Fundamental Viscoelastic Properties of Batter

2.4.1. Dynamic Oscillatory Test

A model batter sample was prepared containing all the ingredients listed in Section 2.2, scaled down to 10 g total. A model batter was freshly prepared for each replicate. The model batter was mixed for 5 min at 100 rpm, using a modified 10-g bowl rotary pin mixer (National Manufacturing, Lincoln, NE, USA). An oscillatory frequency sweep test was based on Hesso et al. (2015) with modifications [17]. An AR-1000N rheometer (TA Instruments, New Castle, DE, USA) was used with a 25-mm plate geometry (cross-hatched to minimize slippage) and a gap of 1 mm at 25 °C. During the test, the plate geometry and batter sample area was covered with a trap to prevent drying or moisture loss. The sample was allowed to recover for 1 min before starting the test. Test conditions included 0.1 to 10 Hz at 0.5% strain, within the linear viscoelastic region. G' (storage modulus), G'' (loss modulus), and tan delta (δ) were determined by this dynamic oscillatory measurement. Analyses were performed in three independent samples.

2.4.2. Temperature-Ramp Test

A non-isothermal method was applied to study gelation of the batter [18–20]. Model batter samples were prepared, as described in Section 2.4.1. Non-isothermal heating of a small amplitude oscillatory shear test (temperature ramp) was measured with an AR-1000N rheometer (TA Instruments, New Castle, DE, USA). A recovering period of 1 min was used before starting the test. Test conditions included a temperature range of 25 to 90 °C, at a heating rate of 5 °C per min, with a constant frequency at 1 Hz and 0.5% strain within the linear viscoelastic region, in which the stress was directly proportional to strain. Analyses were done in three independent samples. Parameters recorded were G' (elastic modulus, Pa), the stored energy in each batter sample or recoverable cycle of deformation and G'' (viscous modulus, Pa), the loss energy in each batter sample or viscous dissipation per cycle of deformation. Complex shear modulus (G^*) was calculated (Equation 1, where i represented an imaginary unit) to record the overall resistance to flow and deformation of the samples. G^* is a useful property as it is a direct measure of the rigidity of soft solid structure of a material, when exposed to stresses below the yield stress [20,21]. High value of G^*, means more solid structure or more resistance to deformation of the samples.

$$G^* = G' + iG' \tag{1}$$

The kinetic parameters of the model (Equation (2)) were obtained by numerical methods, according to Alvarez et al. (2017); it was performed with the numerical derivative of the elastic moduli with respect to time. Where T_0 temperature (K) at the minimum value of dG'/dt, a, b, and c are the constants of the model, and c can be used to calculate the Ea (kJ.mol^{-1}) activation energy (Equation (3)), where R is 8.314 J mol^{-1} K^{-1} (the universal gas constant). Activation energy (Ea) revealed the energy required for gelatinization [14].

$$\frac{dG'}{dt} = a + be^{-c\left(\frac{1}{T} - \frac{1}{T_0}\right)} \tag{2}$$

$$Ea = cR \tag{3}$$

2.5. Firmness of Red-Kidney-Bean–Rice Cupcakes

Cupcake firmness is among the key texture parameters that influence consumer sensory perception and acceptance. After the cupcake samples were cooled to ambient temperature (25 °C), firmness was measured using a TA-XT2 Texture Analyzer (Texture Technologies Corp., Hamilton, MA, USA/Stable Micro Systems, Godalming, Surrey, UK). Analysis was performed following the Approved Method outlined by the American Association of Cereal Chemists (AACC 74-10.02) for bread, using a 25-mm diameter cylindrical probe and 40% compression. Cupcake firmness was performed with 12 independent replicates per sample.

2.6. Sensory Evaluation of Red-Kidney-Bean–Rice Cupcakes

Organoleptic properties of red kidney bean-rice cupcakes were evaluated with a 9-point hedonic scale ranging from 'dislike extremely' (1) to 'like extremely' (9). Cupcakes were tested after being cooled for 60 min. The samples labeled with random 3-digit codes were scored by twenty untrained panelists, students of the Postharvest Technology at Maejo University, Thailand. The sensory attributes included flavor, taste, texture, and overall quality.

2.7. Statistical Analysis

Results are expressed as mean values ± standard deviation of the minimum three independent analyses, unless otherwise described, such as cupcake firmness ($n = 12$) and sensory evaluation ($n = 20$). Mean comparison for significant differences of each treatment was tested by Duncan's New Multiple Range, using the SAS program (Version 9.1 SAS Institute Inc., Cary, NC, USA). Correlations of the rice flour level effect on batter and cupcake properties were estimated with Principal Component Analysis (PCA) by Canoco for the Windows 5 software (Centre for Biometry, Wageningen, The Netherlands).

3. Results and Discussion

3.1. Effect of Rice Flour Addition on Red Kidney Bean Batter Consistency

Consistency was estimated with a 100 g-aliquot of full formula batter by measuring the flow rate in a consistometer, as used in the industry for quality control. The test was simple, fast, and provided enough information for consistency standardization. The levels of rice flour significantly decreased the flow rate (increased consistency) of gluten-free RKB cupcake batter, starting at 10% rice flour addition (Table 1). Rice flour addition of 10%, 15%, and 25% significantly decreased the flow rate of the gluten-free cupcake batter by 4%, 12%, and 20%, respectively, compared to the control sample. Data fitted a second-degree polynomial ($y = -0.0022x^2 - 0.0659x + 14.163$, $R^2 = 0.955$) reflecting negative slopes starting at 10% rice level. Increased consistency of the batter matrix could be explained in part by the increase in air entrainment forming air bubbles, which increase the elastic character of the batter [22]. This might have been favored by rice flour properties and specifically of rice starch. We speculated that the continuous viscous phase of batter containing rice flour has more organized hydrogen bonds with water molecules, resulting in increased consistency, as observed by a decreased flow rate. A study of rice flour and starch showed that water-binding capacity of rice flour was higher than rice starch, which were 128% and 100%, respectively [23]. The value of rice was within the range of red kidney bean flour, with 125% water absorption [24]. Starch damage in the rice flour might also have contributed to an increased water-holding capacity [25,26]. The results agreed with reports of a flow rate study of rice flour and wheat flour, which proposed that a decrease in flow rate in batter was due in part to the rate of flour water absorption [27,28]. Previous studies have showed that amylose content is negatively correlated with water absorption, swelling power, and damaged starch [29–31]. Rice flour has amylose content approximately 15% lower than red kidney bean flour [32,33]. Thus, this low amylose content of rice flour could impact the consistency of batter.

Table 1. Batter and cupcake characteristics of gluten-free red kidney bean flour with rice flour addition [a,b].

Rice Flour Level [c] (%)	Batter Characteristics			Cupcake Characteristic
	Consistency (cm/min)	T_0 (°C)	Ea (kJ mol^{-1})	Firmness (g)
0	14.3 ± 0.3 [a]	63.0 ± 2.5 [a]	165 ± 3 [b]	2970 ± 5 [a]
5	14.0 ± 0.1 [a]	60.2 ± 2.3 [b]	191 ± 2 [a,b]	2767 ± 4 [a]
10	13.2 ± 0.3 [b]	56.8 ± 2.5 [c]	177 ± 2 [b]	1709 ± 3 [b]
15	12.3 ± 0.3 [c]	56.7 ± 2.3 [c]	206 ± 4 [a,b]	1871 ± 3 [b]
25	11.2 ± 0.3 [d]	56.0 ± 2.5 [c]	237 ± 2 [a]	1619 ± 3 [b]

[a] Consistency (flow rate), Temperature ramp test: T_0, inflection temperature of gelation and Ea, activation energy of the batter measured as flow rate. *Firmness*, cake texture. [b] Mean values ($n = 3 ±$ SD) followed by a different letter within a column are significantly different, Duncan test ($p < 0.05$). [c] Rice flour addition to red kidney bean flour.

3.2. Effect of Rice Flour Addition on Red Kidney Bean Batter Viscoelastic Properties

3.2.1. Frequency Sweep Test of Batter

G' (storage modulus) and G'' (loss modulus) increased as the levels of rice flour increased in the frequency range of 0.1 to 10 Hz (Figure 1). Elastic behavior dominated in all gluten-free RKB cupcake batter samples ($G' > G''$). Cake batter is considered a wet foam with contributions to its bulk rheology from the continuous viscous phase and the discontinuous bubble phase (elastic effect) introduced during mixing and the production of gas from leavening agents [22]. At 1 Hz and 0.5% strain, tan δ of the batter with 25% rice addition decreased by 18%, compared to the control (data not shown), suggesting that the solid character (stiffness) of the batter increased by 18%. From Figure 1, a 2.3 fold increase in the solid-like behavior (G') of the batter (22.4 Pa for control up to 50.7 Pa for 25% rice flour) was higher than the 1.3 fold increase of the liquid-like behavior (G'') (17.6 Pa for control and up to 22.1 Pa for 25% rice flour) (Figure 1). Adequate viscosity is a factor contributing to the containment of the bubbles. Structural visualization studies are needed to identify the distribution of the starch (specifically the separation of small and large starch granules), proteins, and fat, as well as the size of the bubbles in the batter. Hesso et al. (2015) reported that air incorporation increased with an increase in batter viscosity [17]. It is also postulated that the rice starch might align in the interphase of oil in the batter [23]. The results are in overall support of the increase in consistency of the full formula batter described in Section 3.1, as it decreased the flow rate after rice addition. In agreement to this study, a report of rice-based gluten-free batter for muffins with different levels of cowpea protein isolates found that the elastic behavior increased more than the viscous behavior [25]. In our study, enhanced viscoelasticity of the RKB gluten-free batter with increased levels of rice flour suggested the formation of composite matrix structures with enhanced water-binding capacity, as mentioned earlier (Section 3.1). We speculated that a decrease in free water availability for other ingredients, mainly protein and fiber that form more disorganized structures, would result in a more organized matrix. The levels of rice flour in the RKB batter system improved the gel rigidity and the network of RKB batter [7]. In addition, the results showed a solid-like behavior (G') for the batter; when the rice flour content varied from 5% to 15%, the rice flour showed a similar behavior.

Figure 2 depicted the log–log plots of G' and G'' that were used to evaluate if the elastic and storage moduli followed a power law relationship in the RKB batter samples, with different rice flour levels. If the response was linear, then the relationship followed a power law, suggesting that a constant increase of G' led to a constant increase in G''. The same plot, known as the Han plot, was used to determine the compatibility of polymer blends [34,35]. Overall, the separation of the curve containing 25% rice flour from the control RKB flour was observed as a shift to higher elasticity (G'), compared to the control ($p = 0.05$) (Figure 2). The tracings of 0% to 15% rice converged, suggesting that they are compatible. At the terminal area, all the curves merge (0% to 25% rice flour), suggesting that at a higher frequency, the lines would have similar correlation. [35]. The 25% rice flour curve separates by shifting to a rise in elasticity compared from the other rice levels, suggesting the formation of more heterogeneous structures compared to the other samples, through the polymers present in the flours [36,37]. The increased values for G' with G' in 25% rice flour compared to other levels might be a contribution of the multiple structures, where less protein and fiber from the RKB flour and overall more starch from the rice flour formed a new continuous network with a higher organization (lower enthalpy), as compared to the control RKB flour and other rice levels. Our previous study of RKB batter with different heat-moisture treatments showed that the high temperature altered the viscoelastic properties of the RKB batter; however, their structures were homogenous [15]. Thus, Figure 2 confirmed the heterogeneous nature of the RKB batter containing rice flour.

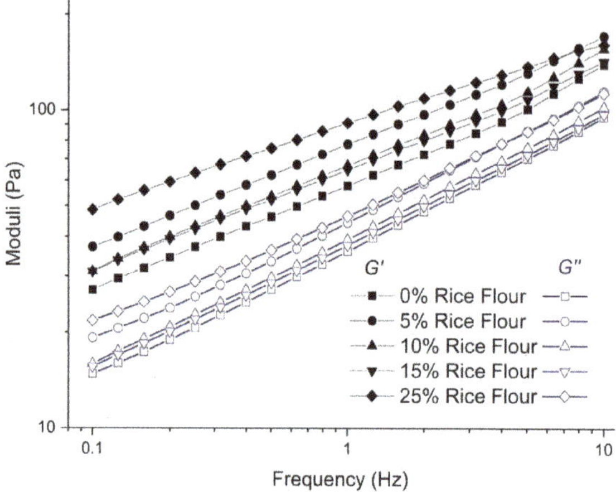

Figure 1. Moduli (G' and G'') as a function of frequency of red kidney bean gluten-free batter with different rice flour levels.

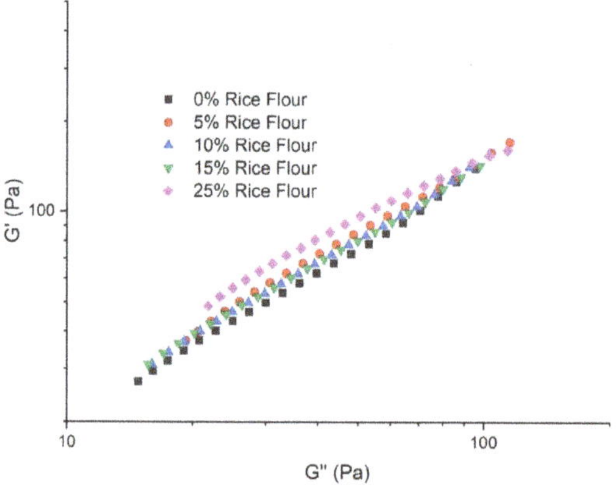

Figure 2. Log–log plot of G' as a function of G'' of red kidney bean gluten-free batter with different rice flour levels.

3.2.2. Temperature Ramp Test of Batter

The viscoelastic properties of the RKB batter with and without rice during heating from 25 °C to 90 °C with constant values of frequency and strain indicated that the complex shear modulus (G^*) increased as the tan δ decreased (Figure 3). The almost linear response of the complex shear modulus changed to an upward curve after the gelation onset, indicating an increase in gel rigidity during heating, which reached up to a 94% increase (at 90 °C) with 25% rice flour, compared to the starting temperature of 25 °C. RKB batter with different rice flour levels showed overall similar rising gel rigidity (G^*) trends after 50 °C (Figure 3) suggesting similar viscoelastic patterns, during gelatinization.

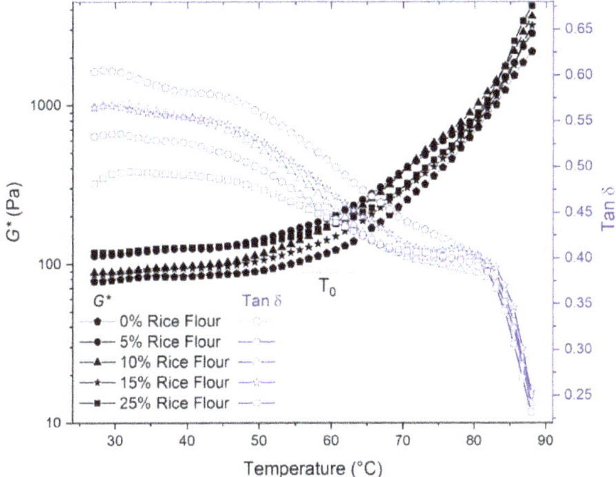

Figure 3. Complex modulus G* (left axis) and tan (δ) (right axis) as a function of the temperature of the red kidney bean gluten-free batter with different rice flour levels. Filled symbols, complex modulus G*; and open symbols tan δ.

The tracing curves of tan δ showed two (rice flour additions) and three (RKB flour) clear inflection points, suggesting different changes of structures (phase transitions) as a function of increasing temperature (Figure 3). Rice flour at 25% had a significantly ($p = 0.05$) lower tan δ at the initial stages of heating before the gelation onset, compared to the rest of the samples, suggesting a higher solid-like character (G') in this sample. That higher solid-like behavior was conserved up to around 50 °C, where the decrease of tan δ accelerated, suggesting a rapid increase in solid-like behavior. This was attributed to the starch and protein hydration, resulting in gel rigidity onset. Tan δ curves crossed-over the G^* curves at around 60 °C (rice flour additions) and 65 °C (RKB flour control), suggesting more rapid changes occurring in the structures. Most likely these changes were related to the denaturation of proteins and the rapid starch granule swelling at those temperatures [29,30]. The predominant formation of hydrogen bonds with water and additional hydrophobic interactions from protein aggregation and starch swelling account for the increase G^* and decrease tan δ (Figure 3), in the continuous phase of the batter.

Activation energy (E_a) was used as an indication of the energy used for the critical gel rigidity during gelatinization, while the inflection temperature (T_0) was used for the point at which the batter viscoelasticity increased at the initial stage of gel point. T_0 is normally associated with batter density and level of gelatinization [14]. Initiation of gelatinization in all RKB batter samples was in the range of 56 to 63 °C (Table 1). These observations agreed with literature reports where onset temperature variation during starch gelatinization from four kidney bean cultivars was in the range of 61.4–66.9 °C, while the onset temperature for rice starch gelatinization was 61 °C, measured by differential scanning calorimetry (DSC) [38,39]. Reported gelatinization range (70 to 85 °C) for basmati rice starch measured by an increase in G' was higher than the values of our study [39]. Increase in rice flour content decreased the RKB batter inflection gelation temperature (T_0) from 63 to 56 °C, with no significant ($p = 0.05$) change in the activation energy (E_a). Significant difference ($p = 0.05$) in activation energy was observed only between the control RKB flour and 25% rice addition, where a 44% increase was needed to reach the critical gel rigidity (Table 1). Activation energy is related to phase transitions during gelatinization of starch, glass transition of starch granules in the amorphous region, and melting transition in the crystalline region [39]. Thus, the addition of rice flour induced gelation at lower temperatures and the process required a higher activation energy, only when rice flour was added at 25% ($p < 0.05$). This could be partially explained by the smaller starch granule size of rice flour (3–8 µm), compared to the kidney bean starch (16–60 µm), increase in overall starch content in the system, and RKB flour protein reduction [32,33]. Literature reports for rice and kidney bean flour gelatinization enthalpy, individually, are in a similar range 7.5–15.6 J/g for rice flour and 10.8–15.4 J/g for kidney bean flour, respectively [40,41]. Activation energy (*Ea*) from this non-isothermal heating of small amplitude oscillatory shear test could be useful to indicate the changes during gelatinization.

3.3. Effect of Rice Flour Addition on Red Kidney Bean Cupcake Firmness

Cupcake texture estimated by firmness is one of the most important parameters for consumer acceptance. As the level of rice flour increased, the RKB cupcakes firmness decreased, reaching a 45.5% reduction with 25% rice flour, compared to the control (Table 1). These observations agreed with a reduced (50% lower) firmness of a rice-based, gluten-free muffin, compared to muffins containing white cowpea protein isolate at 12% level [10]. A decrease in cupcake firmness might be attributed to the factors discussed earlier including the smaller starch granule size of rice, compared to RKB and different molecular organization of rice, lowering the overall protein content in cupcakes, combined with a redistribution of water availability in the gluten-free cupcake. Water absorption and protein content were 2.1 and 2.8 times higher in RKB flour, compared to rice flour (2.6 and 1.23 g/g and 20.5% and 7.4%, for RKB and rice flour, respectively) [15,40]. The incorporation of rice flour changed the structural order of the batter by diluting the proteins and lowering the firmness of gluten-free RKB rice cupcake. Amylose content in rice (average 25%) is lower than kidney bean flour (34% to 41%) and would result in less retrogradation after baking [33,40]. Thus, diluting the amylose content of kidney bean flour by rice flour addition reduced gluten-free cupcake hardness. Previous studies showed that an increase in batter viscosity allowed gas entrapped within the batter structure. This resulted in higher cake volume, crumb with increased airy structure, thinner walls, and thus lower firmness [42,43].

3.4. Effect of Rice Flour Addition on Red Kidney Bean Cupcake Sensory Properties

Rice flour impacted the organoleptic properties of red kidney bean cupcake, with higher sensory scores observed with the samples containing rice flour. The red-kidney-bean–rice cupcakes were scored from 'like slightly' to 'like moderately' (Table 2). There was no change in the sensory scores at the 5% rice flour level. Scores for flavor and taste attributes of the red-kidney-bean–rice cupcakes were higher when rice flour was added at 10%. All attributes had a highest score with 25% rice flour. It was interesting to note that the sensory panelists did not observe changes in texture until 25% rice flour level, when they scored the cupcakes higher, while the overall quality score was improved at 15% rice flour. The nutritional value of the RKB cupcakes of 100 g serving size included 404 Kcal, 18 g total fat, of which 4 g were saturated fat, 104 mg cholesterol, 187 mg sodium, 48 g total carbohydrates, 9 g dietary fiber, 27 g sugar, and 14 g of protein. No significant differences in nutritional value were observed with addition of rice flour at up to 25%.

Table 2. Sensory properties for red-kidney-bean–rice cupcakes [a,b].

Rice Flour Level (%) [b]	Flavor	Taste	Texture	Overall Quality
0	6.1 ± 0.1 [b]	6.2 ± 0.2 [b]	6.2 ± 0.5 [b]	6.1 ± 0.1 [b]
5	6.2 ± 0.2 [b]	6.3 ± 0.1 [b]	6.1 ± 0.2 [b]	6.0 ± 0.2 [b]
10	7.1 ± 0.3 [a]	7.2 ± 0.2 [a]	6.2 ± 0.1 [b]	6.1 ± 0.1 [b]
15	7.3 ± 0.4 [a]	7.1 ± 0.0 [a]	6.1 ± 0.2 [b]	7.1 ± 0.1 [a]
25	7.1 ± 0.3 [a]	7.1 ± 0.2 [a]	7.1 ± 0.1 [a]	7.2 ± 0.2 [a]

[a] Mean values ($n = 20 \pm$ SD) followed by a different letter within a column are significantly different, Duncan test ($p < 0.05$). Evaluated by a 9-point hedonic scale ranging from 'dislike extremely' (1) to 'like extremely' (9). [b] Rice flour level addition to red kidney bean flour.

3.5. Variation of Red Kidney Bean Batter and Cupcake Properties with Rice Flour Addition

Total explained variance from all variables was 92.4% (Figure 4). Principal component 1 (PC1) had the majority of explained variance (80.3%), while the contribution of principal component 2 (PC2) was only 12.1% (Figure 4). The main contributors of PC1 (first coordinate axis) were batter complex shear modulus, activation energy of gelatinization, and tan δ. The main contributor to PC2 was the sample with 10% rice addition—situated almost at the top, at 90 degrees, at the center of the graph. However, only one parameter (*G″* viscous behavior) moved slightly to the horizontal PC2, meaning it had some variations explained by it. The fact that 10% rice addition was almost in a direction opposite to the viscous behavior also suggest that it was inversely correlated to 10% rice addition. However, as the contribution to the variance was low—only 12.1%—compared to 80.3% contribution to the variance of PC1, it was better to highlight the contributions of PC1. Notice how all parameters were along PC1. The vectors aligned in opposite direction revealed negative correlation, while the vectors in the same direction were positively correlated. The results from the loading plot (Figure 4) revealed that cupcake firmness was positively correlated with batter consistency, tan δ (TanDelta in the graph) and inflection temperature (T_0) of gelation, and the treatments associated with those parameters were 0% and 5% rice levels. Complex modulus G*, solid (*G′*) and viscous (*G″*) moduli were positively correlated with the activation energy (*Ea*) of gelatinization and the highest rice flour level (25%), which was the farthest away from the center of the graph. Rice flour addition to RKB flour was negatively correlated to cupcake firmness. These results confirmed that the most variation was observed at the highest rice level (25%). This study also revealed that the gelation of the batter and its consistency were good indicators of quality control for gluten-free RKB–rice flour cupcakes.

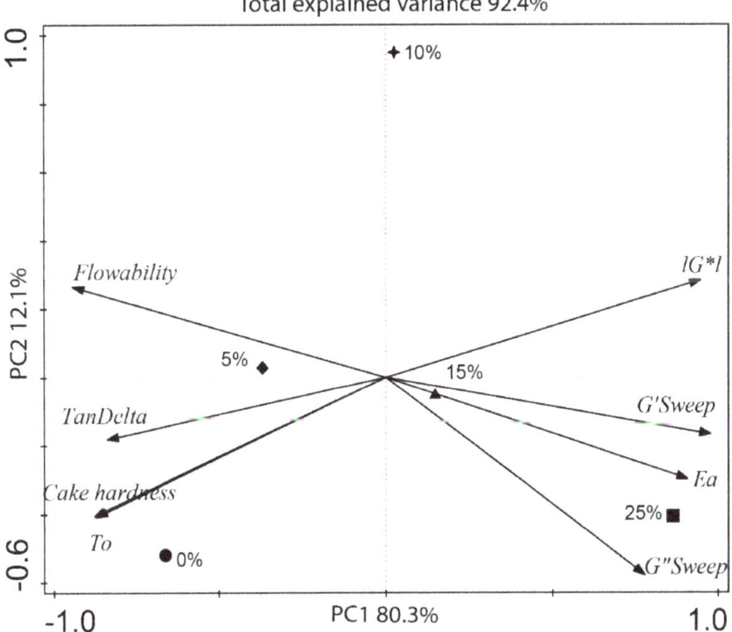

Figure 4. Redundancy analysis of red kidney bean gluten-free batter with and without rice flour with 8 indicators of the viscoelastic properties of gluten (definitions in Table 1), showing the effect of five rice flour levels (0%, 5%, 10%, 15%, and 25%). Symbols: 0%—circle, 5%—diamond, 10%—star, 15%—up triangle, and 25%—square.

4. Conclusions

The level of rice flour addition improved the viscoelastic properties of gluten-free RKB batter and cupcake softness. With 25% of rice flour addition in gluten-free RKB flour, the system increased the batter's solid-like and viscous behavior, batter consistency, inflection of gelatinization and temperature, and produced a softer cupcake texture. The RKB batter with rice followed a power law and had a heterogeneous structure based on the log–log plot of G' vs G''. In addition, the activation energy of gelatinization increased with 25% rice incorporation. This study described an improvement of the RKB batter and cupcake macro-structural characteristics with addition of rice.

Author Contributions: P.C. designed experiments and wrote original draft preparation. N.K. performed research work. Z.J.H.E. developed sample preparation, data analysis and revised manuscript. P.R.-D. acquired funding, supervised the research work, discussed the data, reviewed and edited manuscript. All authors have read and agreed to the published version of the manuscript.

Funding: This research was funded by the Agricultural Research Development Agency (ARDA, Public Organization), Thailand, Oklahoma State University, and the U.S. Department of Agriculture–National Institute of Food and Agriculture (USDA–NIFA).

Conflicts of Interest: The authors declare no conflict of interest.

References

1. Wani, I.A.; Sogi, D.S.; Wani, A.A.; Gill, B.S. Physico-chemical and functional properties of flours from Indian kidney bean (*Phaseolus vulgaris* L.) cultivars. *LWT Food Sci. Technol.* **2013**, *53*, 278–284. [CrossRef]
2. Mishra, P.K.; Tripathi, J.; Gupta, S.; Variyar, P.S. Effect of cooking on aroma profile of red kidney beans (Phaseolus vulgaris) and correlation with sensory quality. *Food Chem.* **2017**, *215*, 401–409. [CrossRef] [PubMed]
3. Jnawali, P.; Kumar, V.; Tanwar, B. Celiac disease: Overview and considerations for development of gluten-free foods. *Food Sci. Hum. Wellness* **2016**, *5*, 169–176. [CrossRef]
4. Vici, G.; Belli, L.; Biondi, M.; Polzonetti, V. Gluten free diet and nutrient deficiencies: A review. *Clin. Nutr.* **2016**, *35*, 1236–1241. [CrossRef]
5. Cornejo, F.; Caceres, P.J.; Martínez-Villaluenga, C.; Rosell, C.M.; Frias, J. Effects of germination on the nutritive value and bioactive compounds of brown rice breads. *Food Chem.* **2015**, *173*, 298–304. [CrossRef]
6. Capriles, V.D.; Arêas, J.A.G. Novel Approaches in Gluten-Free Breadmaking: Interface between Food Science, Nutrition, and Health. *Compr. Rev. Food Sci. Food Saf.* **2014**, *13*, 871–890. [CrossRef]
7. Matos, M.E.; Sanz, T.; Rosell, C.M. Establishing the function of proteins on the rheological and quality properties of rice based gluten free muffins. *Food Hydrocoll.* **2014**, *35*, 150–158. [CrossRef]
8. Torbica, A.; Hadnađev, M.; Dapčević, T. Rheological, textural and sensory properties of gluten-free bread formulations based on rice and buckwheat flour. *Food Hydrocoll.* **2010**, *24*, 626–632. [CrossRef]
9. Shevkani, K.; Singh, N. Influence of kidney bean, field pea and amaranth protein isolates on the characteristics of starch-based gluten-free muffins. *Int. J. Food Sci. Technol.* **2014**, *49*, 2237–2244. [CrossRef]
10. Shevkani, K.; Kaur, A.; Kumar, S.; Singh, N. Cowpea protein isolates: Functional properties and application in gluten-free rice muffins. *LWT Food Sci. Technol.* **2015**, *63*, 927–933. [CrossRef]
11. Fan, J.; Mitchell, J.R.; Blanshard, J.M.V. A model for the oven rise of dough during baking. *J. Food Eng.* **1999**, *41*, 69–77. [CrossRef]
12. Zhang, J.; Datta, A.K. Mathematical modeling of bread baking process. *J. Food Eng.* **2006**, *75*, 78–89. [CrossRef]
13. Narsimhan, G. A mechanistic model for baking of leavened aerated food. *J. Food Eng.* **2014**, *143*, 80–89. [CrossRef]
14. Alvarez, M.D.; Cuesta, F.J.; Herranz, B.; Canet, W. Rheometric Non-Isothermal Gelatinization Kinetics of Chickpea Flour-Based Gluten-Free Muffin Batters with Added Biopolymers. *Foods* **2017**, *6*, 3. [CrossRef]
15. Chompoorat, P.; Rayas-Duarte, P.; Hernández-Estrada, Z.J.; Phetcharat, C.; Khamsee, Y. Effect of heat treatment on rheological properties of red kidney bean gluten free cake batter and its relationship with cupcake quality. *J. Food Sci. Technol.* **2018**, *55*, 4937–4944. [CrossRef]
16. USDA. Food Composition Brand Golden Tropics, Ltd. 2019. Available online: https://fdc.nal.usda.gov/fdc-app.html#/food-details/402558/nutrients (accessed on 5 January 2020).

17. Hesso, N.; Garnier, C.; Loisel, C.; Chevallier, S.; Bouchet, B.; Le-Bail, A. Formulation effect study on batter and cake microstructure: Correlation with rheology and texture. *Food Struct.* **2015**, *5*, 31–41. [CrossRef]

18. Ahmed, J.; Ramaswamy, H.S.; Ayad, A.; Alli, I. Thermal and dynamic rheology of insoluble starch from basmati rice. *Food Hydrocoll.* **2008**, *22*, 278–287. [CrossRef]

19. Yoon, W.; Gunasekaran, S.; Park, J. Characterization of thermorheological behavior of Alaska pollock and Pacific whiting surimi. *J. Food Sci.* **2004**, *69*, 338–343. [CrossRef]

20. Leroy, V.; Pitura, K.M.; Scanlon, M.G.; Page, J.H. The complex shear modulus of dough over a wide frequency range. *J. Non-Newton. Fluid Mech.* **2010**, *165*, 475–478. [CrossRef]

21. Wang, H.; Liu, X.; Zhang, H.; Apostolidis, P.; Erkens, S.; Skarpas, A. Micromechanical modelling of complex shear modulus of crumb rubber modified bitumen. *Mater. Des.* **2020**, *188*, 108467. [CrossRef]

22. Meza, B.E.; Chesterton, A.K.; Verdini, R.A.; Rubiolo, A.C.; Sadd, P.A.; Moggridge, G.D.; Wilson, D.I. Rheological characterisation of cake batters generated by planetary mixing: Comparison between untreated and heat-treated wheat flours. *J. Food Eng.* **2011**, *104*, 592–602. [CrossRef]

23. Falade, K.O.; Christopher, A.S. Physical, functional, pasting and thermal properties of flours and starches of six Nigerian rice cultivars. *Food Hydrocoll.* **2015**, *44*, 478–490. [CrossRef]

24. Chompoorat, P.; Phimphimol, J. Development of a Highly Nnutritional and Functional Gluten Free Cupcake with Red Kidney Bean Flour for Older Adults. *Food Appl. Biosci. J.* **2019**, *7*, 16–26.

25. Kim, J.M.; Shin, M. Effects of particle size distributions of rice flour on the quality of gluten-free rice cupcakes. *LWT Food Sci. Technol.* **2014**, *59*, 526–532. [CrossRef]

26. Vandeputte, G.; Delcour, J. From sucrose to starch granule to starch physical behaviour: A focus on rice starch. *Carbohydr. Polym.* **2004**, *58*, 245–266. [CrossRef]

27. Majzoobi, M.; Imani, B.; Sharifi, S.; Farahnaky, A. The effect of particle size and level of rice bran on the batter and sponge cake properties. *J. Agr. Sci. Tech.* **2013**, *15*, 1175–1184.

28. Mukprasirt, A.; Herald, T.; Flores, R. Rheological characterization of rice flour-based batters. *J. Food Sci.* **2000**, *65*, 1194–1199. [CrossRef]

29. Horstmann, S.W.; Belz, M.C.; Heitmann, M.; Zannini, E.; Arendt, E.K. Fundamental study on the impact of gluten-free starches on the quality of gluten-free model breads. *Foods* **2016**, *5*, 30. [CrossRef]

30. Donovan, J.W. A study of the baking process by differential scanning calorimetry. *J. Sci. Food Agric.* **1977**, *28*, 571–578. [CrossRef]

31. Sasaki, T.; Matsuki, J. Effect of wheat starch structure on swelling power. *Cereal Chem.* **1998**, *75*, 525–529. [CrossRef]

32. Hoover, R.; Hughes, T.; Chung, H.; Liu, Q. Composition, molecular structure, properties, and modification of pulse starches: A review. *Food Res. Int.* **2010**, *43*, 399–413. [CrossRef]

33. Zhou, Z.; Robards, K.; Helliwell, S.; Blanchard, C. Composition and functional properties of rice. *Int. J. Food Sci. Technol.* **2002**, *37*, 849–868. [CrossRef]

34. Walha, F.; Lamnawar, K.; Maazouz, A.; Jaziri, M. Rheological, morphological and mechanical studies of sustainably sourced polymer blends based on poly (lactic acid) and polyamide 11. *Polymers* **2016**, *8*, 61. [CrossRef]

35. Han, C.D.; Kim, J. Rheological technique for determining the order–disorder transition of block copolymers. *J. Polym. Sci. Pol. Phys.* **1987**, *25*, 1741–1764. [CrossRef]

36. Ahmed, J.; Almusallam, A.S.; Al-Salman, F.; AbdulRahman, M.H.; Al-Salem, E. Rheological properties of water insoluble date fiber incorporated wheat flour dough. *LWT Food Sci. Technol.* **2013**, *51*, 409–416. [CrossRef]

37. Ahmed, J.; Auras, R.; Kijchavengkul, T.; Varshney, S.K. Rheological, thermal and structural behavior of poly (ε-caprolactone) and nanoclay blended films. *J. Food Eng.* **2012**, *111*, 580–589. [CrossRef]

38. Hager, A.S.; Arendt, E.K. Influence of hydroxypropylmethylcellulose (HPMC), xanthan gum and their combination on loaf specific volume, crumb hardness and crumb grain characteristics of gluten-free breads based on rice, maize, teff and buckwheat. *Food Hydrocoll.* **2013**, *32*, 195–203. [CrossRef]

39. Rangelov, A.; Arnaudov, L.; Stoyanov, S.; Spassov, T. Gelatinization of industrial starches studied by DSC and TG. *Bulg. Chem. Commun.* **2017**, *49*, 422–429.

40. Fujita, S.; Fujiyama, G. The Study of Melting Temperature and Enthalpy of Starch from Rice, Barley, Wheat, Foxtail and *Proso millets*. *Starch Stärke* **1993**, *45*, 436–441. [CrossRef]

41. Feizollahi, E.; Mirmoghtadaie, L.; Mohammadifar, M.A.; Jazaeri, S.; Hadaegh, H.; Nazari, B.; Lalegani, S. Sensory, digestion, and texture quality of commercial gluten-free bread: Impact of broken rice flour type. *J. Texture Stud.* **2018**, *49*, 395–403. [CrossRef]
42. Shelke, K.; Faubion, J.M.; Hoseney, R.C. The dynamics of cake baking as studied by a combination of viscometry and electrical resistance oven heating. *Cereal Chem.* **1990**, *67*, 575–580.
43. Wilderjans, E.; Pareyt, B.; Goesaert, H.; Brijs, K.; Delcour, J.A. The role of gluten in a pound cake system: A model approach based on gluten–starch blends. *Food Chem.* **2008**, *110*, 909–915. [CrossRef]

 *foods*

Article

Bioactive Compounds, Antioxidant Activity, and Sensory Analysis of Rice-Based Extruded Snacks-Like Fortified with Bean and Carob Fruit Flours

Claudia Arribas [1], Blanca Cabellos [1], Carmen Cuadrado [1], Eva Guillamón [2] and Mercedes M. Pedrosa [1,*]

1   Food Technology Department, SGIT-INIA, Ctra de La Coruña, Km 7.5., 28040 Madrid, Spain
2   Centre for the Food Quality, INIA, C/Universidad s/n, 42004 Soria, Spain
*   Correspondence: mmartin@inia.es

Received: 3 July 2019; Accepted: 27 August 2019; Published: 2 September 2019

Abstract: Generally, extruded gluten-free foods are mostly phytochemically deficient. In this study inositol phosphates, α-galactosides, lectins, protease inhibitors, and phenols, their antioxidant activity and sensorial analysis of some rice/bean/whole carob fruit flour blends were determined in unprocessed (controls) and extruded formulations. The fortification of rice-based extrudates with both legumes has a positive influence on both their bioactive compound content and their acceptability by consumers. The extruded formulations contained around twice as much ($p < 0.05$) total α-galactosides than their unprocessed counterparts. Extrusion significantly reduced the phytic acid content (10%) and significantly increased the less phosphorylated forms (16%–70%). After extrusion, the lectins and protease inhibitors were eliminated. The different phenolic compounds mostly increased (11%–36%), notably in the formulations with carob fruit. The antioxidant activity and the different groups of phenols showed a positive correlation in the extrudates. All the experimental extrudates had higher amounts of bioactive compounds than the commercial extruded rice. Considering the amount of phytochemicals determined in the novel gluten-free extrudates and the scores of sensorial analysis, formulations containing 20%–40% bean and 5% carob fruit could be adequate in promoting health-related functions, helping to increase pulse consumption, and allowing the food industry to satisfy consumers' requirement for functional foods.

Keywords: legumes enrichment; galactosides; phytate; protease inhibitors; phenols

1. Introduction

Nowadays, consumers demand 'free-from' foods to improve their health and wellbeing. Among the free-from foods, gluten-free (GF) products have become very popular, mainly because consumers perceive these GF foods as healthier than the corresponding gluten-containing counterparts. However, it has been reported that a GF diet might not be recommended for the general population due to the nutritional deficiencies associated with GF diets [1].

There are different, naturally whole GF grains that have shown to contain good nutritional quality and good bioactive compound content, which are intrinsically related to healthy benefits. Dehusked rice (*Oryza sativa* L.) is reckoned as the most appropriate ingredient for the formulation of a wide variety of GF foods, such as breakfast cereals, cereal-based snacks, or dietetic foods. However, although whole rice can supply many bioactive compounds, after dehusking and polishing, the obtained white rice loses many phytochemicals, such as phytate, polyphenols, and protease inhibitors [2]. Beans, like other pulses, are well-known sources of phytochemicals (e.g., α-galactosides, phytates, protease inhibitors, phenols, or lectins), which are traditionally considered as antinutritional factors since they can reduce the digestion/absorption of some nutritive compounds or cause intestinal discomfort. Nevertheless, nowadays, some of these compounds are recognised as bioactive compounds that provide beneficial health effects (prebiotic, anti-tumour, or anti-inflammatory effects), playing a pivotal role in reducing the risk or the prevention of heart diseases, some cancers, obesity, type 2 diabetes, and Parkinson's disease [3–6]. However, their antinutrient or pro-nutrient effect has been related to the amount present in foods, as well as to interactions with other compounds present in the diet [4,5]. The fruits of the carob tree (*Ceratonia siliqua* L.), an underutilised leguminous tree from the Mediterranean region, have been traditionally exploited as a feed ingredient. The carob seeds are mostly utilised for the production of locus bean gum or carob bean gum (E-410), which is of great importance for the food industry as a stabiliser or thickener. The seedless pods (named kibble) have mainly been utilised for feed and as a cocoa substitute in cakes, breads, confectionery, or drinks because of its high content of dietary fibre and soluble sugars [7]. Furthermore, carob fruits contain a range of phytochemicals, mainly polyphenols and dietary fibre, as well as low amounts of α-galactosides and inositol phosphates [7,8].

Recent evidence suggests that a GF diet reduced beneficial gut bacteria as a result of the reduced intake of oligosaccharides from wheat (such as oligofructose and inulin) with a prebiotic action [1], therefore, the use of legumes, such as bean and carob fruit, would be of great interest as functional ingredients to formulate pro-health or functional foods, providing bioactive compounds to white-rice based products, and increasing the protein and dietary fibre content in these foods [2]. Moreover, the new food products fortified with legumes would lead to increasing pulse consumption and promote a healthy Mediterranean diet.

Extrusion is a versatile technique used by the industry that enables the formulation of different types of ready-to-eat products such as snacks and breakfast cereals with a high rate of consumption, mainly by young people. They are mainly based in rice or corn flours and/or starches and tend to have low nutritional quality and high calorie content [9,10]. A recommended solution to enhance their nutritional quality is the enrichment of rice flour with legume flours [2]. Different pulses (beans, peas, lentils, chickpeas, soy, etc.) have been treated by extrusion/cooking, though they are generally blended with rice or corn starch to develop good expanded products [11–14]. It has been reported that extrusion is an efficient process to modify the content of some phytochemicals, although with ambivalent results, since the extrusion conditions as well as the raw materials used can increase or reduce the amount of these compounds [3,11,13,15,16], therefore, it is necessary to determine these changes to obtain maximum health benefits in novel extrudates.

Although Ravindran, et al. [17] described the use of carob bean gum to enhance the characteristics of extruded rice-pea formulas, to the best of our knowledge, there is little information [3,10] about the utilisation of carob fruit flours in the preparation of extruded foods. Furthermore, according to the literature consulted, there is no information on the bioactive compound composition of novel GF extrudates based on the linked use of rice, whole carob fruit, and bean flours.

Therefore, the main aim of this research was to study the extrusion effect on some phytochemicals content (α-galactosides, inositol phosphates, lectins, protease inhibitors, and phenolic compounds), as well as the antioxidant activity of various rice-based blends enriched with bean and whole carob fruit flours. The data obtained from the extrudates was compared with their unprocessed equivalents (controls) and with a commercial extruded rice. The results will update the knowledge related to the presence of bioactive compounds in innovative GF products made from legume formulations.

2. Materials and Methods

2.1. Raw Materials and Formulated Flours

The ingredients used for the formulation of the extrudates were white rice (*O. sativa* var. Montsianell), bean (*Phaseolus vulgaris* var. Almonga), and whole carob fruit (*Ceratonia siliqua* var Negreta and Roja). The ingredients were obtained from Cámara Arrocera de Amposta (Tarragona, Spain), a local farmer (Benjamín Rodríguez Álvarez, León, Spain), and Armengol Hermanos (Tarragona, Spain), respectively. The seeds and fruits were powdered using a Retsch SK1 mill (Haan, Germany) fitted with a 1 mm sieve. Different percentages of rice (50%–80%), bean (20% or 40%), and whole carob fruit (5% or 10%) flours were mixed to produce six formulations coded as 20.0, 20.5, 20.10, 40.0, 40.5, and 40.10. The codes used correspond to the following: The first number corresponds to the bean percentage and the second one to the carob fruit percentage included in the blends. The results from the extrudates were compared to an extruded rice sample (100%) obtained in the market.

2.2. Extrusion Process

A Clextral EVOLUM 25 twin-screw extruder (Clextral, Firminy, France) from CARTIF (Valladolid, Spain) was utilised. The extruder was equipped with a 25 mm screw diameter and a 600 mm barrel length. The mixtures were added to the extruder at a rate of 25 kg/h. The last barrel temperature was 125 °C. The screw speed was 900 rpm for 20% bean formulations and 950 rpm for 40% bean formulations. The rate of water addition was 2.5, 3.0, and 3.2 kg/h for samples without carob fruit flour, with 5% carob fruit flour, and with 10% carob fruit flour, respectively. The extrudates were reduced to flour using a Tecator mill (Cyclotec 1093, Höganäs, Sweden) equipped with a 1 mm screen and stored at room temperature in polyethylene bags until the analysis was performed.

2.3. Analysis of Bioactive Compounds

The analysis of raw materials, unprocessed mixtures (as controls), and extruded samples, as well as the commercial extruded sample was performed.

2.3.1. Soluble Sugars and α-Galactosides

Sugars were extracted and then analysed with a HPLC system coupled to a refractive index detector (Beckman System Gold Instrument, Los Angeles, CA, USA) following the method described by Pedrosa, et al. [18]. A total of 20 µL of each sample was injected in a spherisorb-5-NH2 column (250 × 4.6 mm i.d., Waters, Milford, MA, USA) equilibrated with acetonitrile/water (60:40, *v/v*) at a flow rate of 1 mL/min. Each individual sugar was identified and quantified by using the corresponding calibration curves from external standards (Sigma, St. Louis, MO, USA). Ciceritol was obtained from Dr. A. I. Piotrowicz-Cieslak (Olsztyn-Kortowo, Poland). A linear response was observed in the range (0–4 mg/mL), with a correlation coefficient of 0.99. The curves obtained were: Sucrose ($y = 0.193x - 0.001$), galactinol ($y = 0.227x - 0.014$), raffinose ($y = 0.193x + 0.003$), ciceritol ($y = 0.295x + 0.030$), and stachyose ($y = 0.217x - 0.008$). Original HPLC chromatograms of sugars and galactosides of all the analysed samples are presented in supplementary Figure S3.

2.3.2. Inositol Phosphates

Individual inositol phosphates (IP3–IP6) were extracted and determined by HPLC and refractive index detection according to Burbano, et al. [19]. A total of 10 µL of Aliquots were injected on a PRP-1 column (150 × 4.1 mm i.d., 5 µm, Hamilton, Reno, Nevada, USA), and was maintained at 45 °C. The mobile phase was a mixture of methanol/water (52/48, *v/v*) with 0.8% of tetrabutyl ammonium hydroxide (40% in water, Sigma, St. Louis, MO, USA), 0.05% of 91% formic acid (Sigma, St. Louis, MO, USA), 100 µL of a phytic acid hydrolysate (6 mg/mL), and the pH was adjusted to 4.3 with 5 M of sulphuric acid. The flow rate was 1 mL/min. The quantification was developed by using a calibration curve obtained from sodium phytate (0–5 mg/g, $y = 0.144x + 0.016$, $R^2 = 0.99$) (Sigma, St. Louis, MO, USA). Original HPLC chromatograms of IP3-IP6 of all the analysed samples are presented in supplementary Figure S4.

2.3.3. Protease Inhibitors

The trypsin inhibitors (TI) and chymotrypsin inhibitors (CI) were obtained following the method described by Pedrosa, et al. [20]. The trypsin inhibitor units (TIU) per mg of flour were determined as described by Welham and Domoney [21]. One TIU was defined as that which gave a reduction in absorbance units measured at 410 nm of 0.01 relative to the trypsin control reaction in a 10 mL assay volume. Chymotrypsin inhibitor units (CIU) per mg of flour were determined according to Sathe and Salunkhe [22]. One CIU was defined as an increase of 0.01 absorbance units at 256 nm of the reaction mixture relative to the chymotrypsin control reaction.

2.3.4. Lectins

The lectin content in the raw materials was determined using a haemagglutination assay with trypsin-treated rat blood cells (Wistar rats, Animal-housing unit of Complutense University, Madrid, Spain). To quantify the *P. vulgaris* lectin (PHA) in raw bean- and bean-containing formulations, a competitive indirect ELISA assay was carried out according to Pedrosa, Cuadrado, Burbano, Muzquiz, Cabellos, Olmedilla-Alonso, and Asensio-Vegas [20]. One haemagglutination unit (HU) was defined as the amount of material (mg) causing 50% agglutinated erythrocytes. The values were expressed as HU/kg flour for the haemagglutination assay, and as a % PHA for the ELISA assay. In each ELISA assay, two bean cultivars (Procesor and Pinto) were included as positive and negative controls, respectively. The calibration curve of a standard PHA (0–500 ng/mL) used for quantification of bean lectins was $y = -0.4311\mathrm{Ln}(x) + 2.711$ ($R^2 = 0.98$).

2.3.5. Phenolic Compounds

Phenolic compounds were extracted twice using a mixture of methanol-HCl (1000/1)/water (80:20, *v/v*) at room temperature for 16 h [23]. A total of 1 mL of the combined extracts was mixed with 1.5 mL of a solution of 2% HCl in ethanol/water (80/20, *v/v*). These solutions were utilised for the quantification of anthocyanins, flavonoles, tartaric esters, and total phenols by a spectrophotometric method (Beckman DU-640, Los Angeles, CA, USA) that monitored the absorbance at 520, 360, 320, and 280 nm, respectively [24]. The extracts were also utilised for the determination of the antioxidant activity by using the ORAC method [25]. Calibration curves obtained from commercial cyanidin-3-glucoside (C3Glc) (0–20 µg/mL; $y = 0.060x + 0.060$; $R^2 = 0.99$) (Extrasynthèse, Germay, France), quercetin (Q) (0–80 µg/mL; $y = 0.059x + 0.083$; $R^2 = 0.99$), caffeic acid (CA) (0–30 µg/mL $y = 0.095x + 0.009$; $R^2 = 0.99$), and catechin (C) (0–200 µg/mL $y = 0.012x + 0.005$; $R^2 = 0.99$) (Aldrich, Munich, Germany) were utilised to quantify anthocyanins, flavonols, tartaric esters, and total phenols, respectively, according to Oomah, et al. [24].

2.4. Sensory Evaluation

The 6 extrudates (Ex-20.0, Ex-20.5, Ex-20.10, Ex-40.0, Ex-40.5, and Ex-40.10) were evaluated by 10 semi-trained panellists. Samples were codified with random numbers, placed in small plastic boxes, and given to the panellists. They were instructed to cleanse their mouth with water between each sample evaluated and they were provided with written instructions. Extrudates were evaluated for colour, odour (intensity, roasted, cooked pulse, nuts, unpleasant odours), flavour (intensity, salty, pulses, unpleasant flavours), texture (crunchiness, hardness, adhesivity, fracture, chewiness), and overall quality, according to a nine-point scale where 9 = extremely like or high intensity and 1 = extremely dislike or low intensity).

2.5. Statistical Analysis

A representative and homogeneous amount (25 kg) of each formulation was split into in two parts and then extruded. The chemical analyses of raw materials, unprocessed, and extruded samples and the commercial sample were carried out in quadruplicate. Data are reported as mean ± standard error. A prior analysis of the normality and homogeneity of variance of all variables was performed using Shapiro–Wilks and Levenes's test, respectively. One-way ANOVA and a Duncan's multiple range test were used to analyse data at the 95% confidence level ($p < 0.05$). Pearson's correlation coefficients between the different bioactive compounds and the ORAC values were determined. Moreover, a principal component analysis (PCA) of the standardised data of the measured variables was performed to ensure their reliability and to obtain meaningful interpretations of the results (Supplementary Materials). All the statistical analyses were developed using Statgraphics Centurion XVII.II software (Graphics Software System, Rockville, MD, USA).

3. Results and Discussion

3.1. Effect of Bean and Whole Carob Flour Fortification on the Content of Some Bioactive Compounds of Rice-Based Unprocessed Formulations

As observed in Tables 1–4, raw bean flour showed the highest content of α-galactosides (raffinose and stachyose), inositol phosphates (IP3–IP6), and protease inhibitors. Raw carob fruit flour contained the lowest amount of total inositol phosphates but also the highest sucrose, total phenols, and antioxidant activity. Regarding the haemagglutination activity measured using trypsin-treated rat blood cells (data not shown), raw bean flour showed the highest lectin content (10 HU/kg), followed by carob fruit (0.32 HU/kg), and rice (0.16 HU/kg). Moreover, raw rice contained low amounts of sucrose and TI. The bioactive compound content established in the raw materials was close to that reported in the literature [3,4,25,26].

Table 1. Effect of extrusion treatment on the soluble sugars, ciceritol, and α-galactosides (mg/g dry weight) content of raw materials, the non-extruded (NE-) and the extruded (Ex-) flour formulations, and the commercial sample.

Sample	Sucrose	Galactinol	Raffinose	Ciceritol	Stachyose	Total α-Galactosides
Bean	30.00 ± 0.95 [e,f,g]	2.29 ± 0.05 [a,b]	5.92 ± 0.09 [c]	0.34 ± 0.01 [a]	26.85 ± 0.25 [g]	32.77 ± 0.25 [g]
Carob fruit	150.46 ± 10.04 [h]	n.d. *	5.84 ± 0.02 [c]	n.d.	n.d.	5.84 ± 0.02 [a]
Rice	2.98 ± 0.15 [a]	n.d.	n.d.	n.d.	n.d.	n.d.
NE-20.0 **	8.65 ± 0.15 [a,b,A]	2.53 ± 0.10 [a,b,c,A]	2.84 ± 0.13 [a,A]	0.78 ± 0.04 [b,A]	8.55 ± 0.40 [a,A]	11.39 ± 0.55 [b,A]
NE-20.5	16.85 ± 0.29 [c,d,A]	2.97 ± 0.06 [b,c,d,e,A]	3.08 ± 0.03 [a,A]	0.90 ± 0.02 [b,c,A]	7.59 ± 0.17 [a,A]	10.67 ± 0.15 [b,A]
NE-20.10	29.91 ± 0.96 [g,h,A]	3.30 ± 0.07 [c,d,e,A]	3.14 ± 0.04 [a,A]	0.94 ± 0.03 [b,c,d,A]	8.61 ± 0.40 [a,A]	11.76 ± 0.45 [b,A]
NE-40.0	12.44 ± 0.15 [c,A]	3.58 ± 0.03 [d,e,A]	3.87 ± 0.04 [b,A]	1.14 ± 0.05 [c,d,e,A]	11.82 ± 0.23 [c,A]	15.69 ± 0.25 [c,A]
NE-40.5	22.83 ± 0.73 [d,e,f,A]	3.11 ± 0.13 [b,c,d,e,A]	3.97 ± 0.08 [b,A]	1.19 ± 0.05 [d,e,A]	13.37 ± 0.40 [d,A]	17.34 ± 0.49 [d,A]
NE-40.10	30.06 ± 0.41 [h,A]	1.87 ± 0.09 [a,A]	3.89 ± 0.09 [b,A]	1.25 ± 0.04 [e,A]	12.37 ± 0.28 [c,d,A]	16.26 ± 0.48 [c,d,A]
Ex-20.0	10.27 ± 0.08 [a,b,A]	2.84 ± 0.11 [b,c,d,A]	9.46 ± 0.35 [e,B]	5.44 ± 0.07 [f,B]	10.30 ± 0.43 [b,B]	19.76 ± 0.78 [e,B]
Ex-20.5	22.10 ± 0.19 [d,e,B]	10.27 ± 0.34 [h,B]	7.60 ± 0.27 [d,B]	5.64 ± 0.15 [f,B]	11.51 ± 0.50 [c,B]	19.10 ± 0.81 [e,B]
Ex-20.10	34.62 ± 0.10 [e,f,g,A]	9.71 ± 0.44 [h,B]	8.12 ± 0.26 [d,B]	5.42 ± 0.07 [f,B]	13.36 ± 0.07 [d,B]	21.48 ± 0.33 [f,B]
Ex-40.0	17.84 ± 0.16 [c,d,A]	3.81 ± 0.10 [e,A]	12.21 ± 0.24 [f,B]	8.92 ± 0.07 [h,B]	20.91 ± 0.13 [e,B]	33.13 ± 0.50 [g,h,B]
Ex-40.5	31.64 ± 0.05 [f,g,A]	5.31 ± 0.25 [f,B]	12.16 ± 0.15 [f,B]	8.93 ± 0.13 [h,B]	22.14 ± 0.20 [f,B]	34.30 ± 0.28 [h,B]
Ex-40.10	41.60 ± 0.07 [e,f,g,B]	7.31 ± 0.33 [g,B]	12.05 ± 0.34 [f,B]	8.41 ± 0.01 [g,B]	20.98 ± 0.20 [e,f,B]	33.03 ± 0.14 [g,h,B]
Commercial extruded rice	71.40 ± 1.31 [g]	n.d.	n.d.	n.d.	n.d.	n.d.
p value	<0.0001	<0.001	<0.0001	<0.001	<0.0001	<0.001

* n.d. not detected. Values are mean ± standard error ($n = 4$); mean values in the same column followed by a different superscript are significantly ($p < 0.05$) different; small superscript letters mean differences between all the samples analysed, whereas capital superscript letters mean differences due to extrusion treatment for the same formulation. ** Sample codes: 20.0 (20% bean; 0% whole carob fruit); 20.5 (20% bean; 5% whole carob fruit); 20.10 (20% bean; 10% whole carob fruit); 40.0 (40% bean; 0% whole carob fruit); 40.5 (40% bean; 5% whole carob fruit); and 40.10 (40% bean; 10% whole carob fruit).

To study the extrusion effect on the phytochemical content of the novel formulations, the unprocessed (NE-) formulations were used as controls. The rice/legume blends are a complex matrix that can affect the extraction of bioactive compounds [12,18,20], therefore, it is necessary to know the amount of different phytochemicals in the unprocessed samples.

The addition of whole carob fruit and bean flour had a positive impact on the phytochemical composition of non-extruded (NE-) mixtures. The higher legume percentage produced more α-galactosides (Table 1) and inositol phosphates (Table 2) content in the NE-blends, with stachyose and phytic acid (IP6) being the main compounds detected. The amount of sugars in the NE-blends showed, in general, the following pattern: Sucrose > stachyose > raffinose > galactinol > ciceritol. The total amount of galactosides, on average, was 11.27 mg/g in the blends with 20% bean and 16.43 mg/g in the blends with 40% bean. A considerable increase in sucrose (Table 1) and phenolic compounds (Table 4) content was detected in the NE-samples containing carob fruit flour (NE-20.5, NE-20.10,

NE-40.5, and NE-40.10) due to the high content of both compounds determined in raw whole carob fruit. In relation to the protease (trypsin and chymotrypsin) inhibitor content (Table 3), the lowest activities corresponded to the formulations with 20% bean and the highest values coincided with the 40% blends. The addition of carob fruit flour did not produce significant changes ($p > 0.05$) in the protease inhibitor activities analysed. The content of lectin (PHA) in the different NE-formulations, evaluated by competitive ELISA did not show significant ($p > 0.05$) differences, ranging from 0.035% to 0.108% PHA. As carob fruit contained higher amounts of phenols than bean and rice, the levels of the different phenolic compounds in NE-blends increased as a function of increasing the proportion of carob. The same tendency was found for the ORAC. A positive correlation was observed among total phenols and ORAC ($R^2 = 0.58$, $p < 0.05$) and flavonols and ORAC ($R^2 = 0.59$, $p < 0.05$). Moreover, the incorporation of carob fruit and bean in the formulation of different rice-based blends had a positive impact on the bioactive compound content. This was also observed by other authors who added different legumes (navy bean, kidney bean, small red bean, pea, faba bean, or chickpea) to unprocessed formulations based on starch or flour from corn or rice [2,11,14,25,27].

Table 2. Effect of extrusion treatment on the inositol phosphate content (mg/g dry weight) of raw materials, the non-extruded (NE-) and the extruded (Ex-) flour formulations, and the commercial sample.

Sample	IP3	IP4	IP5	IP6	Total Inositol Phosphates
Bean	0.26 ± 0.01 [e]	0.42 ± 0.01 [e]	1.39 ± 0.03 [g]	10.12 ± 0.03 [j]	12.20 ± 0.04 [h]
Carob	n.d. *	0.15 ± 0.01 [b]	0.36 ± 0.04 [b]	0.15 ± 0.01 [a]	0.66 ± 0.03 [a]
Rice	0.10 ± 0.01 [b,c]	0.03 ± 0.03 [a]	0.22 ± 0.01 [a]	1.53 ± 0.05 [b]	1.88 ± 0.03 [b]
NE-20.0 **	0.22 ± 0.01 [d,e,A]	0.24 ± 0.01 [c,A]	0.53 ± 0.02 [c,A]	3.32 ± 0.11 [e,f,A]	4.31 ± 0.12 [f,g,A]
NE-20.5	0.22 ± 0.01 [d,e]	0.25 ± 0.01 [c,A]	0.53 ± 0.01 [c,A]	3.28 ± 0.12 [e,A]	4.27 ± 0.10 [f,A]
NE-20.10	0.22 ± 0.01 [d,e]	0.27 ± 0.01 [c,A]	0.57 ± 0.01 [c,A]	3.88 ± 0.01 [h,i,A]	4.93 ± 0.02 [h,A]
NE-40.0	0.22 ± 0.01 [d,e,A]	0.27 ± 0.01 [c,A]	0.71 ± 0.01 [d,A]	4.08 ± 0.06 [i,j,A]	5.28 ± 0.07 [i,A]
NE-40.5	0.22 ± 0.01 [d,e,A]	0.27 ± 0.01 [c,A]	0.75 ± 0.01 [d,A]	4.18 ± 0.09 [h,A]	5.42 ± 0.10 [i,A]
NE-40.10	0.22 ± 0.01 [d,e,A]	0.24 ± 0.01 [c,A]	0.90 ± 0.03 [e,A]	4.54 ± 0.04 [i,A]	5.90 ± 0.07 [j,A]
Ex-20.0	0.05 ± 0.01 [a,b,B]	0.29 ± 0.01 [c,A]	0.69 ± 0.02 [d,B]	2.61 ± 0.03 [c,B]	3.65 ± 0.09 [d,B]
Ex-20.5	n.d.	0.29 ± 0.01 [c,A]	0.71 ± 0.02 [d,B]	2.98 ± 0.11 [d,B]	3.98 ± 0.14 [e,B]
Ex-20.10	n.d.	0.27 ± 0.01 [c,A]	0.74 ± 0.01 [d,B]	3.56 ± 0.12 [f,g,B]	4.56 ± 0.16 [g,B]
Ex-40.0	0.05 ± 0.01 [a,b,B]	0.37 ± 0.02 [d,e,B]	1.20 ± 0.05 [e,B]	3.89 ± 0.03 [g,h,B]	5.51 ± 0.08 [i,A]
Ex-40.5	0.05 ± 0.01 [a,b,B]	0.41 ± 0.01 [e,B]	1.21 ± 0.06 [e,f,B]	3.74 ± 0.06 [g,h,B]	5.41 ± 0.06 [i,A]
Ex-40.10	0.16 ± 0.01 [c,d,A]	0.35 ± 0.01 [d,B]	1.27 ± 0.02 [f,B]	4.33 ± 0.14 [h,i,A]	6.11 ± 0.15 [j,A]
Commercial extruded rice	0.12 ± 0.01 [b,c]	0.30 ± 0.01 [c]	0.75 ± 0.01 [d]	1.01 ± 0.05 [b]	2.25 ± 0.03 [c]
p value	<0.0001	<0.0001	<0.0001	<0.0001	<0.0001

* n.d. not detected. Values are mean ± standard error ($n = 4$); mean values in the same column followed by a different superscript are significantly ($p < 0.05$) different; small superscript letters mean differences between all the samples analysed, whereas capital superscript letters mean differences due to extrusion treatment for the same formulation. ** Sample codes: 20.0 (20% bean; 0% whole carob fruit); 20.5 (20% bean; 5% whole carob fruit); 20.10 (20% bean; 10% whole carob fruit); 40.0 (40% bean; 0% whole carob fruit); 40.5 (40% bean; 5% whole carob fruit); and 40.10 (40% bean; 10% whole carob fruit).

Table 3. Effect of extrusion treatment on trypsin inhibitors (TIU/mg dry weight), chymotrypsin inhibitors (CIU/mg dry weight), and lectin content (%PHA *) in raw materials, the non-extruded (NE-) and the extruded (Ex-) flour formulations, and the commercial sample.

Sample	Trypsin Inhibitors	Chymotrypsin Inhibitors	Lectins
Bean	23.21 ± 0.66 [e]	7.74 ± 0.28 [d]	0.297 ± 0.012 [C]
Carob fruit	0.30 ± 0.02 [a]	n.d. **	n.d.
Rice	0.15 ± 0.01 [a]	n.d.	n.d.
NE-20.0 ***	4.10 ± 0.09 [b,A]	1.97 ± 0.09 [a,b,A]	0.035 ± 0.002 [a]
NE-20.5	4.16 ± 0.06 [b,A]	1.44 ± 0.08 [a,A]	0.045 ± 0.002 [a,b]
NE-20.10	5.53 ± 0.20 [c,A]	1.66 ± 0.08 [a,b,A]	0.052 ± 0.002 [a,b]
NE-40.0	7.83 ± 0.11 [d,B]	5.65 ± 0.17 [c,B]	0.108 ± 0.005 [a,b]
NE-40.5	7.73 ± 0.30 [d,B]	5.93 ± 0.28 [c,B]	0.103 ± 0.005 [b]
NE-40.10	7.34 ± 0.29 [d,B]	5.76 ± 0.25 [c,B]	0.101 ± 0.005 [a,b]
Ex-20.0	n.d.	n.d.	n.d.
Ex-20.5	n.d.	n.d.	n.d.
Ex-20.10	n.d.	n.d.	n.d.
Ex-40.0	n.d.	n.d.	n.d.
Ex-40.5	n.d.	n.d.	n.d.
Ex-40.10	n.d.	n.d.	n.d.
Commercial extruded rice	0.09 ± 0.01 [a]	n.d.	n.d.
p value	<0.0001	<0.0001	<0.0001

* PHA = *Phaseolus vulgaris* lectin. ** n.d. not detected. Values are mean ± standard error ($n = 4$); mean values in the same column followed by a different superscript are significantly ($p < 0.05$) different; small superscript letters mean differences between all the samples analysed, whereas capital superscript letters mean differences due to extrusion treatment for the same formulation. *** Sample codes: 20.0 (20% bean; 0% whole carob fruit); 20.5 (20% bean; 5% whole carob fruit); 20.10 (20% bean; 10% whole carob fruit); 40.0 (40% bean; 0% whole carob fruit); 40.5 (40% bean; 5% whole carob fruit); and 40.10 (40% bean; 10% whole carob fruit).

Table 4. Effect of extrusion on the anthocyanins (µg C3GlcE */g dry weight), flavonols (µg QE/g dry weight), tartaric esters (mg CAE/g dry weight), and total phenols (mg (+) CE/g dry weight) content and antioxidant activity (µmoles Trolox/g dry weight) of raw materials, the non-extruded (NE-) and the extruded (Ex-) flour formulations, and the commercial sample.

Sample	Anthocyanins	Flavonols	Tartaric Esters	Total Phenols	Antioxidant Activity (ORAC)
Bean	36.96 ± 0.24 [i]	0.08 ± 0.001 [c]	0.21 ± 0.01 [i]	2.88 ± 0.02 [g]	24.33 ± 0.07 [i]
Carob fruit	18.00 ± 0.15 [e]	0.75 ± 0.001 [b,c]	0.72 ± 0.01 [b,c]	20.73 ± 0.10 [i]	69.89 ± 1.62 [j]
Rice	18.70 ± 0.83 [f]	0.03 ± 0.001 [b]	0.02 ± 0.001 [a]	0.90 ± 0.03 [a,b]	3.80 ± 0.30 [a]
NE-20.0 **	10.32 ± 0.07 [a]	0.02 ± 0.001 [a]	0.06 ± 0.001 [b,A]	0.71 ± 0.03 [a,A]	8.35 ± 0.04 [b]
NE-20.5	11.00 ± 0.03 [a,b,A]	0.04 ± 0.001 [a,b,c]	0.09 ± 0.001 [c,d,A]	1.38 ± 0.03 [c, A]	9,66 ± 0.45 [c,d,e]
NE-20.10	12.42 ± 0.23 [a,A]	0.07 ± 0.001 [b,c]	0.11 ± 0.001 [e,f,A]	2.26 ± 0.01 [e,f,A]	10.22 ± 0.50 [d,e]
NE-40.0	15.23 ± 0.24 [c]	0.03 ± 0.001 [a,b,A]	0.09 ± 0.001 [c,d,A]	1.10 ± 0.03 [b,A]	9.41 ± 0.45 [b,c,d]
NE-40.5	15.45 ± 0.36 [c,d,A]	0.06 ± 0.001 [a,b,c]	0.11 ± 0.001 [e,f,A]	1.80 ± 0.03 [d,A]	11.57 ± 0.50 [f,g,h]
NE-40.10	17.08 ± 0.36 [e,A]	0.08 ± 0.001 [c]	0.14 ± 0.001 [g,h,A]	2.35 ± 0.02 [e,f,A]	12,36 ± 0.58 [h,A]
Ex-20.0	12.41 ± 0.52 [h]	0.02 ± 0.001 [a]	0.07 ± 0.001 [b,B]	0.92 ± 0.03 [b,B]	8.92 ± 0.38 [b,c]
Ex-20.5	16.78 ± 0.31 [d,e,B]	0.05 ± 0.001 [a,b,c]	0.10± 0.001 [d,e,B]	2.16 ± 0.01 [e,B]	9.32 ± 0.42 [b,c,d]
Ex-20.10	17.13 ± 0.13 [e,B]	0.08 ± 0.001 [b,c]	0.14 ± 0.001 [g,B]	3.25 ± 0.06 [h,B]	10.29 ± 0.30 [d,e]
Ex-40.0	18.13 ± 0.32 [e]	0.08 ± 0.001 [c,B]	0.08 ± 0.001 [b,c,B]	1.33 ± 0.04 [c,B]	10.49 ± 0.49 [d,e,f]
Ex-40.5	21.19 ± 0.09 [f,B]	0.06 ± 0.001 [a,b,c]	0.13 ± 0.001 [f,g,B]	2.37 ± 0.11 [f,B]	11.89 ± 0.48 [g,h]
Ex-40.10	23.30 ± 0.65 [g,B]	0.08 ± 0.001 [c]	0.16 ± 0.001 [h,B]	3.17 ± 0.01 [h,B]	10.86 ± 0.50 [e,f,g,B]
Commercial extruded rice	15.59 ± 0.25 [e,f]	0.06 ± 0.001 [a,b,c]	0.14 ± 0.001 [g]	1.96 ± 0.10 [e]	8.81 ± 0.18 [b,c]
p value	<0.0001	<0.001	<0.0001	<0.0001	<0.0001

* C3GlcE (cyanidin-3-glucoside equivalents); QE (quercitin equivalents); CAE (caffeic acid equivalents); and CE ((+) catechin equivalents). Values are mean ± standard error ($n = 4$); mean values in the same column followed by a different superscript are significantly ($p < 0.05$) different; small superscript letters mean differences between all the samples analysed, whereas capital superscript letters mean differences due to extrusion treatment for the same formulation. ** Sample codes: 20.0 (20% bean; 0% whole carob fruit); 20.5 (20% bean; 5% whole carob fruit); 20.10 (20% bean; 10% whole carob fruit); 40.0 (40% bean; 0% whole carob fruit); 40.5 (40% bean; 5% whole carob fruit); and 40.10 (40% bean; 10% whole carob fruit).

3.2. Extrusion Cooking Impact on the Bioactive Compounds of Rice-Based Formulations

3.2.1. Soluble Sugars and α-Galactosides

The amount of soluble sugars and α-galactosides in the different experimental extrudates and the commercial extruded rice is presented in Table 1. Similar to the unprocessed (NE-) blends, the main sugars found in all the extrudates were sucrose, raffinose, and stachyose. In general, the sugar content pattern in the extrudates was the same as in the NE-counterparts. Extrusion produced an increase ($p < 0.05$) in all the analysed sugars. Regarding the total α-galactosides content, the extrudates with 20% (Ex-20) and 40% (Ex-40) bean presented, on average, 1.78 and 2.03-folds more α-galactosides than their equivalent NE-counterpart (NE-20 and NE-40), respectively. This increase could mostly be attributed to the release of carbohydrates linked to other macromolecules as well as to a mechanical alteration of the structure of the food matrix during extrusion, including cell wall damage, which increased the porosity, thus enhancing the diffusion of the extracting solvent within the food matrix, and subsequently improving the sugar extraction [12,18]. These results were similar to those described for extruded pea and lentil flours [3,12]. However, a reduction in the sugar content of extruded legumes has also been reported [28–30] although these authors reported that the extrusion effect on the galactosides content depends on the moisture and temperature of the extrusion process, as well as the seed and/or the formulation used in the development of the extruded products.

The commercial sample did not contain α-galactosides. Compared to the commercial sample, extrudates without carob fruit flour contained between four and seven times less sucrose, while the extrudates containing 5%–10% carob flour showed, on average, between 1.72 (Ex-40.10) and 3.23 (Ex-20.5) times more sucrose because of the high sucrose amount provided by this raw material.

From a physiological perspective, sugars of the raffinose family are recognised as accounting for flatulence associated with legume consumption. However, different studies have reported that these galactosides can be considered as prebiotics [4,5,31]. Their presence in the diet has been associated with prebiotic activity, increasing the bifidobacterias population in the human colon, stimulating the immune system, and reducing diarrhoea or constipation. Moreover, the colonic flora can ferment these sugars, producing a mixture of short fatty acid that reduces glycaemic symptoms and cholesterol once assimilated into the intestine [5,6]. Therefore, the increase in the total amount of α-galactosides after extrusion can be considered as an added-value attribute of the extrudates and allows bean and carob fruit be taken into account as an added-value component in the extruded formulations.

3.2.2. Inositol Phosphates

It is known that inositol hexakisphosphate, phytate, or phytic acid (IP6) form complexes with some minerals (iron, zinc, or calcium), which negatively affects their absorption. However, the lower phosphorylated forms (IP–IP4) are considered to have a significant function in human health, boosting the assimilation of minerals, preventing the formation of kidney stones, and performing key roles in some disorders, such as type 2 diabetes, some types of cancer, and irritable bowel syndrome [4–6,32].

As observed in the NE-blends, IP6 was the principal inositol phosphate detected in all the extrudates (Table 2). Extrusion cooking reduced ($p < 0.05$), on average, 10% of the IP6 content, leading to less phosphorylated forms [3,12,32]. After extrusion, IP4 and IP5 showed a significant increase (16%–52% and 30%–70%, respectively) compared to their corresponding NE-equivalents. A reduction ($p < 0.05$) in the total IP content was also found (7%–15%) for the Ex-20 samples compared to their respective unprocessed formulations, mostly due to the IP6 reduction. These results were in accordance with those described for extruded faba bean, pea, kidney bean, and chickpea [11,25,29,30]. Although IP6 was reduced (5%–10%) in the Ex-40 samples, the total IP content of these samples showed a slight increase (3%–4%), which did not differ significantly from their NE-40 blends, similar to the results of Lombardi-Boccia, et al. [33] for extruded bean, faba bean, chickpea, and lentil. The extent of the changes on the inositol phosphate content is dependent on the raw materials and/or the composition of the formulations, as well as the extrusion conditions used in the formulation of the extrudates [14,30,32].

Regarding these results, extrusion might be considered a good processing technology to reduce IP6 content by retaining the advantageous lower phosphorylated forms in the innovative extrudates.

The commercial extruded rice contained 1.6 to 2.7 times less total IP and 3–5 times less IP6 than the Ex-formulations. This could be due to inositol phosphates (in the experimental extrudates) that were mainly provided by the bean flour.

3.2.3. Protease Inhibitors and Lectins

As shown in Table 3, the extrusion process eliminated TI and CI activities, as well as lectin content in all the extruded blends, which corroborated the heat-sensitive nature of both phytochemicals. Similarly, several authors found a reduction of TI and lectins by 100% in extruded corn/bean, pea, chickpea, kidney bean, and faba bean [11,12,25,27,34]. The commercial extruded rice showed a small TI activity (0.09 TIU/mg), while CI activity and lectins were not detected. It has been reported that protease inhibitors and lectins are key parameters in establishing food quality [4,26,32], since they hamper protein digestion and the assimilation of nutrients, respectively. Therefore, the absence of both compounds in all the extrudates may improve their nutritional quality [25,26,32]. Moreover, as PHA is considered a toxic compound (causing vomiting, bloating, or diarrhoea in humans) [32], from the perspective of lectin toxicity, these results demonstrate that all the extrudates are safe.

3.2.4. Phenolic Compounds and Antioxidant Activity

Phenolic compounds are considered to be natural antioxidants that can extend the shelf-life of food products [30,35]. The phenolic compound (anthocyanins, flavonols, tartaric esters, and total phenols) content and the antioxidant activity (ORAC) of the extrudates is shown in Table 4. The extrusion process affected the studied phenolic groups to a different extent. In general, the Ex-formulation showed a significant increase in the tartaric esters (on average 11%), anthocyanins (24%), and total phenols (36%) content compared to their NE-counterpart, while the flavonols content in the extrudates did not vary significantly from their NE-formulations. The observed increase could be related to the lack of effluents during extrusion, unlike traditional processing (soaking and/or cooking or autoclaving) that prevent the leaching of water-soluble phenols [20,30]. As expected, the highest increases in total phenols (30%–56%) corresponded to the extrudates that included carob bean flour in their composition (Ex-20.5, Ex-20.10, Ex-40.5, and Ex-40.10). The carob fruit pods are high in phenolic compounds bound to the dietary fibres and the extrusion could lead to the release of some non-available phenols bound to the cell wall, thus growing their extractability [3,15].

Regarding the ORAC, the greatest values were found in the extrudates containing 40% bean and 5%–10% carob fruit. The majority of the extruded formulations showed a slight increase (on average 5.4%) in ORAC, albeit, in general, the differences between the ORAC values of each corresponding pair of NE/Ex-formulations were not significant ($p > 0.05$). A positive correlation was determined among anthocyanins, tartaric esters, and ORAC ($R^2 = 0.63$ and 0.58, $p < 0.05$, respectively). However, the antioxidant activity is conditioned by the type of compound and its amount in the sample [24]. The increase found in ORAC values after extrusion might also be attributable to the existence of other food components (e.g., proteins or aromatic compounds) or to the creation of Maillard reaction products throughout the extrusion that contribute to the antioxidant activity analysed [3,11,27].

As reported by different authors [11,14,15,34], extrusion can produce rises or decreases in both the phenol content and the ORAC of many pulses, and they concluded that extrusion at moderately low moisture (<14%) and temperatures below 120 °C retained higher amounts of phenols and showed greater antioxidant activity.

Compared to the commercial sample, the Ex-20.10, Ex-40.5, and Ex-40.10 formulations presented a higher ($p < 0.05$) amount of total phenols (approximately 1.5 times) and anthocyanins (around 1.6 times), as well as a higher ORAC value (from 1.2 times to 1.4 times).

Various studies reported health benefits associated with different phenols and their antioxidant capacity, such as a lower risk for inflammatory processes or colon cancer [16,23,35]. Therefore, considering the results for the phenolic compounds content and antioxidant capacity, the fortification of rice-based extrudates with both whole carob fruit and bean would be helpful in developing healthy foods.

A PCA was performed to classify and characterise the Ex- and NE- formulations according to their phytochemical content. The PCA on the standardised phytochemical data of the NE- and Ex- samples showed that three principal components explained 86.85% of the total variance (supplementary file, Figures S1 and S2, Tables S1 and S2). Total galactosides and the different phenolic compounds had more weight in the characterisation of the samples by PC1, whereas the thermo-labile compounds (lectins and protease inhibitors) had more weight in the characterisation of the samples by PC2. The variance of PC3 was mainly explained by the ORAC.

3.3. Sensory Evaluation of the Extrudates

In the sensory analysis, the panellists determined how each extrudate compared with the other by considering a number of different attributes related to colour, odour, flavour, and texture. The results of the sensory analysis are shown in Figure 1. The higher scores represent better acceptance of the attribute evaluated.

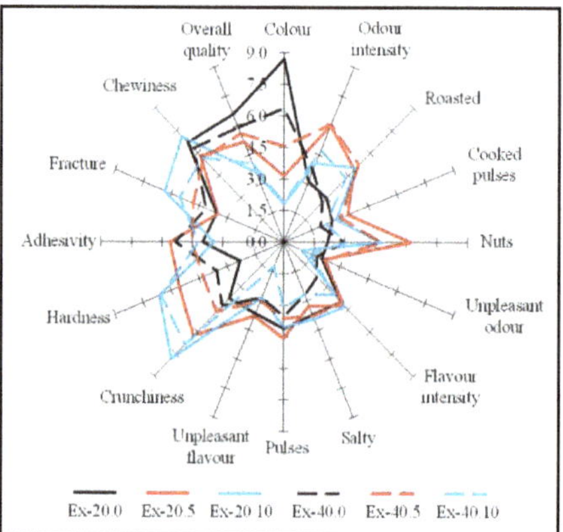

Figure 1. Sensory analysis of the six different extrudates formulated with different proportions of rice, bean, and carob fruit flours. Sample code: Ex-20.0 (20% bean; 0% whole carob fruit); Ex-20.5 (20% bean; 5% whole carob fruit); Ex-20.10 (20% bean; 10% whole carob fruit); Ex-40.0 (40% bean; 0% whole carob fruit); Ex-40.5 (40% bean; 5% whole carob fruit); and Ex-40.10 (40% bean; 10% whole carob fruit).

Textural attributes, mainly crunchiness and hardness, are largely determined to attest to the quality and the consumer acceptability of extruded products, such as snacks or breakfast cereals. The extrudates including carob fruit in their composition showed better textural scores with high crunchiness and low adhesiveness. Crunchiness results were positively correlated ($R^2 = 0.55$, $p < 0.001$) with the presence of carob flour in the formulation.

The colour scores were affected ($p < 0.05$) by the presence of carob flour. A higher amount of carob flour caused a lower preferred colour by the panellists. A strong negative correlation between colour and the presence of carob flour ($R^2 = -0.77$, $p = 0.001$) was found. This could be related to the brown colour of the products formulated with carob flour, leading to a slight rejection of these products. The analysis of the scores of the different aroma and taste parameters tested did not show any correlation with the amount of legumes in the extrudates and no unpleasant odours or flavours were detected.

Regarding the overall quality results, the amount of bean (20% or 40%) in the formulations did not affect ($p > 0.05$) the overall quality of the extrudates. The extrudates containing carob flour showed lower scores, however, the differences between them were not significant ($p > 0.05$). The extrudates elaborated only with rice and bean showed the highest scores, although the addition of 5% carob fruit flour did not significantly affect the overall quality of these extrudates, and therefore these novel extrudates would be well accepted by consumers.

4. Conclusions

From the results obtained, it can be seen that the fortification of rice-based GF puffed products with carob fruit and bean flours had a positive impact on their bioactive compound content. The fortification with carob fruit flour improved their textural attributes and did not significantly affect their overall quality.

The extrusion process affected the studied phytochemicals to a different extent. While total α-galactosides and phenols increased, the IP6 was reduced, and the lectins and protease inhibitors were eliminated. The content of bioactive compounds present in these extrudates might be enough to promote health-associated functions. Moreover, the absence of lectins and protease inhibitors enhanced the nutritional quality of the extrudates. Compared to the commercial extruded rice, all the experimental extrudates showed a higher content of bioactive compounds. Therefore, the formulation of these products would be of interest to both health-conscious consumers and the food industry, allowing it to meet consumers' requirement for functional foods with acceptable sensory attributes.

Supplementary Materials: The following are available online at http://www.mdpi.com/2304-8158/8/9/381/s1. Table S1. Values of the Principal Components (PC1, PC2, and PC3) coefficients for each formulation. Table S2. Component weight of each bioactive compound in relation to the Principal Components (PC1, PC2, and PC3). Figure S1. Principal component analysis (PCA) projection of the two first principal components. NE- (non-extruded formulations). Ex- (extruded formulations). Parameters: Total inositol phosphates (Total IP), sucrose, total galactosides, trypsin Inhibitors (TIU), chymotrypsin inhibitors (CIU), lectins (PHA), anthocyanins, flavonols, tartaric esters, total phenols, and ORAC. Figure S2. Principal component analysis (PCA) projection of the first and third principal components. NE- (non-extruded formulations). Ex- (extruded formulations). Parameters: Total inositol phosphates (Total IP), sucrose, total galactosides, trypsin Inhibitors (TIU), chymotrypsin inhibitors (CIU), lectins (PHA), anthocyanins, flavonols, tartaric esters, total phenols, and ORAC. Figure S3. Original HPLC Chromatograms of sucrose, galactinol, raffinose, ciceritol, and stachyose of the three raw materials, the six non-extruded (NE-), the six extruded (Ex-) flour formulations, and the extruded commercial sample. Figure S4. Original HPLC Chromatograms of inositol phosphates (IP3, IP4, IP5, and IP6) of the three raw samples, the six non-extruded (NE-), the six Extruded (EX-) flour formulations, and the extruded commercial sample.

Author Contributions: M.M.P. conceived and designed research experiments. C.A., B.C., C.C., E.G., and M.M.P. performed the experiments. C.A., B.C., C.C., E.G., and M.M.P. analysed the data. C.A., B.C., C.C., E.G., and M.M.P. wrote, reviewed, and edited the paper. The manuscript was critically revised and approved by all authors.

Funding: The Spanish Ministry of Economy and Competitiveness (Project RTA2012-00042-C02) supported this work. C.A. was supported by a contract (CPR2014-0068) from INIA, and co-financed by the European Regional Development Fund (FEDER) and the European Social Fund (FSE).

Conflicts of Interest: The authors declare no conflict of interest.

References

1. Gaesser, G.A.; Angadi, S.S. Gluten-free diet: Imprudent dietary advice for the general population? *J. Acad. Nutr. Diet.* **2012**, *112*, 1330–1333. [CrossRef] [PubMed]
2. Dalbhagat, C.G.; Mahato, D.K.; Mishra, H.N. Effect of extrusion processing on physicochemical, functional and nutritional characteristics of rice and rice-based products: A review. *Trends Food Sci. Technol.* **2019**, *85*, 226–240. [CrossRef]
3. Arribas, C.; Cabellos, B.; Cuadrado, C.; Guillamón, E.; Pedrosa, M. The effect of extrusion on the bioactive compounds and antioxidant capacity of novel gluten-free expanded products based on carob fruit, pea and rice blends. *Innov. Food Sci. Emerg. Technol.* **2019**, *52*, 100–107. [CrossRef]
4. Muzquiz, M.; Varela, A.; Burbano, C.; Cuadrado, C.; Guillamon, E.; Pedrosa, M.M. Bioactive compounds in legumes: Pronutritive and antinutritive actions. Implications for nutrition and health. *Phytochem. Rev.* **2012**, *11*, 227–244. [CrossRef]

5. Lajolo, F.M.; Genovese, M.I.; Pryme, I.F.; Dale, T.M. Beneficial (antiproliferative) effects of different substances. In *Recent Advances of Research in Antinutritional Factors in Legume Seeds and Oilseeds*; Muzquiz, M., Hill, G.D., Cuadrado, C., Pedrosa, M.M., Burbano, C., Eds.; EAAP Publication: Toledo, Spain, 2004; pp. 123–137.

6. Singh, B.; Singh, J.P.; Shevkani, K.; Singh, N.; Haur, A. Bioactive constituents in pulses and their health benefits. *J. Food Sci. Technol.* **2017**, *54*, 858–870. [CrossRef] [PubMed]

7. Goulas, V.; Stylos, E.; Chatziathanasiadou, M.V.; Mavromoustakos, T.; Tzakos, A.G. Functional Components of Carob Fruit: Linking the Chemical and Biological Space. *Int. J. Mol. Sci.* **2016**, *17*, 1875. [CrossRef]

8. Nasar-Abbas, S.M.; Huma, Z.-E.; Vu, T.-H.; Khan, M.K.; Esbenshade, H.; Jayasena, V. Carob Kibble: A Bioactive-Rich Food Ingredient. *Compr. Rev. Food Sci. Food Saf.* **2016**, *15*, 63–72. [CrossRef]

9. Bouvier, J.M.; Campanella, O.H. Extrusion Technology and Process Intensification. In *Extrusion Processing Technology*; John Wiley & Sons, Ltd.: Hoboken, NJ, USA, 2014; pp. 465–506.

10. Arribas, C.; Cabellos, B.; Sanchez, C.; Cuadrado, C.; Guillamon, E.; Pedrosa, M.M. The impact of extrusion on the nutritional composition, dietary fiber and in vitro digestibility of gluten-free snacks based on rice, pea and carob flour blends. *Food Funct.* **2017**, *8*, 3654–3663. [CrossRef]

11. Anton, A.A.; Gary Fulcher, R.; Arntfield, S.D. Physical and nutritional impact of fortification of corn starch-based extruded snacks with common bean (*Phaseolus vulgaris* L.) flour: Effects of bean addition and extrusion cooking. *Food Chem.* **2009**, *113*, 989–996. [CrossRef]

12. Morales, P.; Berrios, J.; Varela, A.; Burbano, C.; Cuadrado, C.; Muzquiz, M.; Pedrosa, M.M. Novel fiber-rich lentil flours as snack-type functional foods: An extrusion cooking effect on bioactive compounds. *Food Funct.* **2015**, *6*, 3135–3143. [CrossRef]

13. Berrios, J. Extrusion Cooking: Legume Pulses. In *Encyclopedia of Agricultural, Food, and Biological Engineering*, 2nd ed.; CRC Press: Boca Raton, FL, USA, 2010; pp. 453–464.

14. Carvalho, A.; de Andrade Mattietto, R.; Bassinello, P.; Nakamoto Koakuzu, S.; Rios, A.; Maciel, R.; Nunes Carvalho, R. Processing and characterization of extruded breakfast meal formulated with broken rice and bean flour. *Food Sci. Technol.* **2012**, *32*, 515–524. [CrossRef]

15. Brennan, C.; Brennan, M.; Derbyshire, E.; Tiwari, B.K. Effects of extrusion on the polyphenols, vitamins and antioxidant activity of foods. *Trends Food Sci. Technol.* **2011**, *22*, 570–575. [CrossRef]

16. Morales, P.; Cebadera-Miranda, L.; Cámara, R.M.; Reis, F.S.; Barros, L.; Berrios, J.; Ferreira, I.C.F.R.; Cámara, M. Lentil flour formulations to develop new snack-type products by extrusion processing: Phytochemicals and antioxidant capacity. *J. Funct. Foods* **2015**, *19 Pt A*, 537–544. [CrossRef]

17. Ravindran, G.; Carr, A.; Hardacre, A. A comparative study of the effects of three galactomannans on the functionality of extruded pea–rice blends. *Food Chem.* **2011**, *124*, 1620–1626. [CrossRef]

18. Pedrosa, M.M.; Cuadrado, C.; Burbano, C.; Allaf, K.; Haddad, J.; Gelencsér, E.; Takács, K.; Guillamón, E.; Muzquiz, M. Effect of instant controlled pressure drop on the oligosaccharides, inositol phosphates, trypsin inhibitors and lectins contents of different legumes. *Food Chem.* **2012**, *131*, 862–868. [CrossRef]

19. Burbano, C.; Muzquiz, M.; Ayet, G.; Cuadrado, C.; Pedrosa, M.M. Evaluation of antinutritional factors of selected varieties of *Phaseolus vulgaris*. *J. Sci. Food Agric.* **1999**, *79*, 1468–1472. [CrossRef]

20. Pedrosa, M.M.; Cuadrado, C.; Burbano, C.; Muzquiz, M.; Cabellos, B.; Olmedilla-Alonso, B.; Asensio-Vegas, C. Effects of industrial canning on the proximate composition, bioactive compounds contents and nutritional profile of two Spanish common dry beans (*Phaseolus vulgaris* L.). *Food Chem.* **2015**, *166*, 68–75. [CrossRef] [PubMed]

21. Welham, T.; Domoney, C. Temporal and spatial activity of a promoter from a pea enzyme inhibitor gene and its exploitation for seed quality improvement. *Plant Sci.* **2000**, *159*, 289–299. [CrossRef]

22. Sathe, S.K.; Salunkhe, D.K. Studies on trypsin and chymotrypsin inhibitory activities, hemagglutinating activity, and sugars in the great northern beans (*Phaseolus vulgaris* L). *J. Food Sci.* **1981**, *46*, 626–629. [CrossRef]

23. Dueñas, M.; Fernández, M.L.; Hernández, T.; Estrella, I.; Muñoz, R. Bioactive phenolic compounds of cowpeas (*Vigna sinensis* L.). Modifications by fermentation with natural microflora and with *Lactobacillus plantarum* ATCC 14917. *J. Sci. Food Agric.* **2005**, *85*, 297–304.

24. Oomah, B.D.; Cardador-Martínez, A.; Loarca-Piña, G. Phenolics and antioxidative activities in common beans (*Phaseolus vulgaris* L.). *J. Sci. Food Agric.* **2005**, *85*, 935–942. [CrossRef]

25. Alonso, R.; Aguirre, A.; Marzo, F. Effect of extrusion and traditional processing methods on antinutrients and in vitro digestibility of protein and starch in faba and kidney beans. *Food Chem.* **2000**, *68*, 159–165. [CrossRef]

26. Roy, F.; Boye, J.I.; Simpson, B.K. Bioactive proteins and peptides in pulse crops: Pea, chickpea and lentil. *Food Res. Int.* **2010**, *43*, 432–442. [CrossRef]

27. Delgado-Licon, E.; Ayala, A.L.; Rocha-Guzman, N.E.; Gallegos-Infante, J.A.; Atienzo-Lazos, M.; Drzewiecki, J.; Martinez-Sanchez, C.E.; Gorinstein, S. Influence of extrusion on the bioactive compounds and the antioxidant capacity of the bean/corn mixtures. *Int. J. Food Sci. Nutr.* **2009**, *60*, 522–532. [CrossRef] [PubMed]

28. Berrios, J.; Pan, J. Evaluation of extruded black bean (*Phaseolus vulgaris* L.) processed under different screw speeds and particle sizes. In Proceedings of the Annual Meeting of the Institute of Food Technologist, New Orleans, LA, USA, 23–27 June 2001; p. 30.

29. Varela, A.; Guillamón, E.; Cuadrado, C.; Marzo, F.; Burbano, C.; Muzquiz, M.; Pedrosa, M.M. Changes in nutritional active factors and protein in different legumes by extrusion/cooking. In Proceedings of the 6th European Conference on Grain Legumes, Lisboa, Portugal, 16–17 November 2007; p. 56.

30. Pedrosa, M.; Cuadrado, C. The nutritional and Nutraceutical Values od Some ready-to-Eat Expanded Legume Products. In *Legumes for Global Food Secutiry*; Clemente, A., Jimenez-Lopez, J.C., Eds.; Nova Science, Inc.: New York, NY, USA, 2017; pp. 157–182.

31. Tosh, S.M.; Yada, S. Dietary fibres in pulse seeds and fractions: Characterization, functional attributes, and applications. *Food Res. Int.* **2010**, *43*, 450–460. [CrossRef]

32. Nikmaram, N.; Leong, S.Y.; Koubaa, M.; Zhu, Z.; Barba, F.J.; Greiner, R.; Oey, I.; Roohinejad, S. Effect of extrusion on the anti-nutritional factors of food products: An overview. *Food Control* **2017**, *79*, 62–73. [CrossRef]

33. Lombardi-Boccia, G.; Lullo, G.D.; Carnovale, E. In-vitro iron dialysability from legumes: Influence of phytate and extrusion cooking. *J. Sci. Food Agric.* **1991**, *55*, 599–605. [CrossRef]

34. El-Hady, E.A.; Habiba, R.A. Effect of soaking and extrusion conditions on antinutrients and protein digestibility of legume seeds. *LWT Food Sci. Technol.* **2003**, *36*, 285–293. [CrossRef]

35. Aguilera, Y.; Dueñas, M.; Estrella, I.; Hernandez, T.; Benitez, V.; Esteban, R.M.; Martin-Cabrejas, M.A. Phenolic profile and antioxidant capacity of chickpeas (*Cicer arietinum* L.) as affected by a dehydration process. *Plant Foods Hum. Nutr.* **2011**, *66*, 187–195. [CrossRef]

 *foods*

Article

Study of the Effects Induced by Ball Milling Treatment on Different Types of Hydrocolloids in a Corn Starch–Rice Flour System

Luca Nuvoli [1], **Paola Conte** [2,*], **Sebastiano Garroni** [1], **Valeria Farina** [1], **Antonio Piga** [2] **and Costantino Fadda** [2]

[1] Department of Chemistry and Pharmacy, University of Sassari and INSTM, Via Vienna 2, 07100 Sassari, Italy; lunuvoli@uniss.it (L.N.); sgarroni@uniss.it (S.G.); valefari89@gmail.com (V.F.)

[2] Department of Agriculture, University of Sassari, Viale Italia 39/A, 07100 Sassari, Italy; pigaa@uniss.it (A.P.); cfadda@uniss.it (C.F.)

* Correspondence: pconte@uniss.it; Tel.: +39-079-229-277

Received: 20 March 2020; Accepted: 16 April 2020; Published: 20 April 2020

Abstract: The effects of ball milling treatment on both the structure and properties of guar gum (GG), tara gum (TG), and methylcellulose (MC) were analyzed prior to assessing their potential interactions with starch components when they are used alone or in blends in a corn starch–rice flour system. X-ray diffraction profiles showed that the ball milling caused a reduction in the crystallin domain and, in turn, a diminished viscosity of the GG aqueous solutions. Despite an increase in its viscosity properties, effects on TG were minimal, while the milled MC exhibited reduced crystallinity, but similar viscosity. When both milled and un-milled hydrocolloids were individually added to the starch–flour system, the pasting properties of the resulting mixtures seemed to be affected by the type of hydrocolloid added rather than the structural changes induced by the treatment. All hydrocolloids increased the peak viscosity of the binary blends (especially pure GG), but only milled and un-milled MC showed values of setback and final viscosity similar to those of the individual starch. Ball milling seemed to be more effective when two combined hydrocolloids (milled GG and MC) were simultaneously used. No significant differences were observed in the viscoelastic properties of the blends, except for un-milled GG/starch, milled TG/starch, and milled MC/milled TG/starch gels.

Keywords: ball milling; hydrocolloids; starch–flour system; X-ray diffraction; pasting profile; viscoelastic properties

---

1. Introduction

The use of starch and its ability to gelatinize make it a necessary ingredient in food systems [1]. Starch is a macroconstituent of many foods and it is composed of amylopectin and amylose. The organization of the crystalline lamellae within granules is due to the packing of amylopectin into crystallites and this phenomenon is influenced by amylose. The amylopectin structure and the role of amylose are important for the water swelling of starch granules, the thermal properties and gel formation (in water solution) [2]. However, when starch is subjected to processes, such as cooking, shear stress, and cooling, it tends to exhibit synersis, retrogradation, and breakdown, suggesting that its use as a main ingredient in food may be difficult [3]. To overcome these issues, hydrophilic colloids are often used to modify the functionality of starch [4]. Hydrophilic colloids, commonly known as "hydrocolloids", are substances consisting of long chains that have a high molecular weight and an affinity for water, which, in a water-based system, generate gels or highly viscous suspensions. Typically, hydrocolloids are polysaccharides, composed of single or multiple units of sugar, repeated over the entire length of the polymer. Their properties are then influenced by external units, such as sulfate groups, methyl ether groups or carboxyl groups, linked to the main chain [5,6]. The ability to form networks, once dispersed in a solution, and their biocompatibility, makes them excellent additives and/or substituents in food chemistry for the control of microstructure, texture, flavor, and shelf-life [7,8]. Among others, the naturally occurring polysaccharides, such as galactomannans, as well as the chemically synthesized cellulose derivatives, have been widely used in the food industry for their suitable functional properties. In particular, the cellulose derivative methylcellulose has many applications in food systems such as an emulsifier and texturing agent, as well as a thickener and gelling additive [9], whereas galactomannans, such as guar gum and tara gum, are well known to have thickening, binding, and stabilizing abilities [10]. These two galactomannans, however, despite a very similar molecular structure—composed of a β-(1–4)-D-mannan backbone with a single d-galactose branch linked α-(1–6)—differ from each other in the molar ratio of mannose to galactose (3:1 for tara and 2:1 for guar gum), which is responsible for the significant changes in their viscosity and solubility properties, as well as in their interactions with other polysaccharides [10]. The authors of [3] explored the interactions of rice starch with guar gam, hydroxypropylmethylcellulose (HPMC), and xanthan gum to improve the pasting, viscoelastic, and swelling properties of starch–hydrocolloid mixtures. To determine the network stability, the effects of repeated heating–cooling cycles were also analyzed. They found that the tested rice starch–hydrocolloid mixtures (8%, *w/w*) exhibited different pasting, viscoelastic, and swelling properties depending on the type of hydrocolloid added, as well as the concentration levels (0.2%–0.8% *w/w*) used. The authors of [11], when investigating the swelling and pasting properties of non-waxy rice starch–hydrocolloid mixtures using both commercial and laboratory-generated hydrocolloids, observed a strong dependence between the hydrocolloid concentration and swelling power of the rice starch–hydrocolloid blends. They found that the swelling behavior was depressed at low concentrations of hydrocolloids (0%–0.05%), but it increased linearly with increasing hydrocolloid concentrations (0.05%–0.1%). Other hydrocolloids (e.g., alginate, k-carrageenan, xanthan) were investigated by [12] with two different techniques. Recently, it was demonstrated that the use of mechanical treatments, such as ball milling, can alter the functionality of hydrocolloids, influencing, in turn, their interactions with the starch or its constituents [13–15]. Changes in hydrocolloid functionality are probably due to the changes induced by the milling treatment on their structural conformation, thermal properties, particle size distributions, and rheological properties, enhancing the amorphization of the system at the expense of the crystalline structure. The authors of [16] investigated the effects of ball milling on the structural, thermal, and rheological properties of oat bran protein flour. They observed modifications in the structural conformation of both protein and starch as a consequence of the reduction of the particle size and the suppression of the amylose–lipid complex helical structure. The authors of [17] proved that the cellulose crystallite size decreased as a function of increasing milling time.

It has been recognized that the blending of hydrocolloids and cereal starches of various origin, apart from the main above-mentioned standard applications, also plays a critical role in improving both the structure and texture of yeast-leavened baked products for celiac patients. In such products, in which starch is the main component, two of the most used ingredients, due to their wide availability, low price, and desirable rheological properties, are corn and rice (both flour and starch) [7]. In this context, with the aim to achieve suitable functional properties in the development of optimized gluten-free products, it was decided to use rice flour and corn starch as basic ingredients of the starch–flour system.

Starting from these points, the objectives of this study were to explore the interactions of three different hydrocolloids, namely, guar gum, tara gum, and methylcellulose, with starch components when they are used alone or in blends in a corn starch–rice flour system and to analyze the effect of the ball milling treatment on the structure and viscosity of these different hydrocolloids. With this aim, X-ray diffraction (XRD) experiments were carried out to evaluate the potential changes in the conformational structure induced by the ball milling treatment, while the gelling capacity test was conducted to determine potential changes in the viscosity of hydrocolloid aqueous solutions. Viscometric and small deformation rheological measurements were conducted in binary and ternary starch–hydrocolloid mixtures and were compared with the basic starch–flour system (used as a reference sample) to study the potential changes in hydrocolloid functionality in such a complex system.

2. Materials and Methods

2.1. Raw Materials

Rice flour, corn starch, and guar gum (GG) were provided by Chimab (Chimab Food Ingredient Solutions, Campodarsego, PD, Italy); tara gum (TG), under its commercial name AGLUMIX 01, was obtained from Silvateam Food Ingredients (Silvateam Food Ingredients S.r.l., San Michele Mondovì, CN, Italy); methylcellulose (MC) was purchased from Sigma–Aldrich.

2.2. Ball Milling Treatment

Ball milling was performed using a Mixer/Mill 8000 (SPEX SamplePrep, Metuchen, NJ, USA) with a zirconia jar and two zirconia balls (2 g each); 15 g of hydrocolloids was introduced in the vial and milled for 1 h at 875 rpm under air.

2.3. X-ray Diffraction Experiments

The X-ray diffraction patterns of both milled and un-milled hydrocolloids were collected using a SmartLab X-ray powder diffractometer (Rigaku, Tokyo, Japan) aligned according to Bragg–Brentano geometry with Cu Kα radiation (λ = 1.54178 Å) and a graphite monochromator in the diffracted beam. Since the polymers showed broad haloes typical of an amorphous condition, the reciprocal space investigation was deliberately restricted in the angular range from 5° to 60° in 2θ, which allows us to determine the main shape features of patterns.

2.4. Fourier Transform Infrared Analysis

Fourier transform infrared (FTIR) analysis was performed on the examined hydrocolloids, both before and after milling, using an infrared Vertex 70 interferometer (Bruker, Billerica, MA, USA). All samples were ground with ultra-dry KBr (potassium bromide), at a ratio of 1:100, and then pressed into the shape of pellets by applying a uniaxial pressure using a hydraulic press. The spectra were recorded on the hydrogels after drying at 80 °C for 24 h, in transmission mode, in the 400–4000 cm^{-1} range by averaging 64 scans with 4 cm^{-1} resolution. The background was evaluated by measuring the signals of dried air.

2.5. Gelling Capacity of Hydrocolloids

The gelling properties of both milled and un-milled hydrocolloids were determined by a Rapid Visco Analyzer (RVA-4, Newport Scientific, Warriewood, NSW, Australia) using the official method RVA 41.02. Before staring the test, both guar and tara gums (0.28 g) were thoroughly dispersed in 28 mL of distilled water at room temperature, while the methylcellulose aqueous solution was prepared by dispersing the powder in hot water (80 °C) until complete solubilization was reached. The cooling profile used to describe the gelling/thickening behavior of the obtained hydrocolloid aqueous solutions was as follows: samples were held at 80 °C for 5 min, gradually cooled to 20 °C at a rate of 1 °C/min, and finally kept at 20 °C for 5 min. All experiments were performed in duplicate.

2.6. Starch/Flour–Hydrocolloid Blend Preparation

To verify changes in the hydrocolloid functionality induced by the ball milling treatment, both milled and un-milled hydrocolloids were added to a starch-based system composed of corn starch and rice flour (50:50). To this end, twelve different hydrocolloid–starch combinations, which included six binaries and six ternary mixtures, were obtained (Table 1). Firstly, each milled and un-milled hydrocolloid was individually added to the basic starch–flour system to obtain six binary mixtures; secondly, blends of two hydrocolloids, which were obtained by combining the gelling agent methylcellulose (both milled and un-milled) with each of the two thickening agents (both milled and un-milled), were added to the basic starch–flour system to obtain six ternary mixtures. The corn starch–rice flour system alone was used as a reference sample (Table 1).

Table 1. Formulations used to prepare the starch/flour-hydrocolloid blends.

Samples	Corn Starch (g)	Rice Flour (g)	Guar Gum (g)		Tara Gum (g)		Methylcellulose (g)		Water (mL)
			UN	M	UN	M	UN	M	
Corn starch–rice flour	1.75	1.75	-	-	-	-	-	-	25
Binary mixtures									
starch–flour + UN-GG	1.75	1.75	0.25	-	-	-	-	-	25
starch–flour +M-GG	1.75	1.75	-	0.25	-	-	-	-	25
starch–flour + UN-TG	1.75	1.75	-	-	0.25	-	-	-	25
starch–flour + M-TG	1.75	1.75	-	-	-	0.25	-	-	25
starch–flour + UN-MC	1.75	1.75	-	-	-	-	0.25	-	25
starch–flour + M-MC	1.75	1.75	-	-	-	-	-	0.25	25
Ternary mixtures									
starch–flour + UN-MC + UN-GG	1.75	1.75	0.125	-	-	-	0.125	-	25
starch–flour + UN-MC + M-GG	1.75	1.75	-	0.125	-	-	0.125	-	25
starch–flour + M-MC + M-GG	1.75	1.75	-	0.125	-	-	-	0.125	25
starch–flour + UN-MC + UN-TG	1.75	1.75	-	-	0.125	-	0.125	-	25
starch–flour + UN-MC + M-TG	1.75	1.75	-	-	-	0.125	0.125	-	25
starch–flour + M-MC + M-TG	1.75	1.75	-	-	-	0.125	-	0.125	25

UN: un-milled; M: milled; GG: guar gum; TG: tara gum; MC: methylcellulose.

To maintain a constant concentration of starch in both reference and experimental samples, thus allowing comparison among them, the starch–hydrocolloid composites were obtained by addition. The blending of a single hydrocolloid and starch–flour was prepared by firstly suspending the hydrocolloid powder (0.25 g) in 25 mL of distilled water at room temperature and under magnetic stirring, before adding 3.5 g of corn starch–rice flour (50:50). The ternary mixtures were prepared in the same way but, in this case, both hydrocolloids (0.125 g + 0.125 g) were simultaneously suspended in the distilled water before adding the starch powder (Table 1). On the contrary, for all the mixtures containing MC, the hydrocolloid solutions were prepared by dispersing the powder in hot water (80 °C) until a complete solubilization of the particles was achieved; then, the obtained hydrocolloid aqueous solutions were cooled to room temperature and subsequently mixed with the corn starch–rice flour powder. Two different batches of each mixture were prepared, and all the measurements were performed in duplicate.

2.7. Pasting Properties of Starch/Flour–Hydrocolloid Blends

Viscosity profiles of the starch–flour system alone and the starch/flour-hydrocolloid aqueous solutions were determined by a Rapid Visco Analyzer (RVA-4, Newport Scientific, Warriewood, NSW, Australia) according to the International Association of Cereal Science and Technology (ICC) Standard method 162. Before starting the test, the slurries were heated at 50°C and stirred for 5 min at a rotation speed of 960 rpm to achieve complete homogenization. Then, the slurries were held at 50 °C for 1 min, gradually heated to 95 °C, held at this temperature for 2 min 30 s, and finally cooled to 50 °C and held at this temperature for 2 min. The pasting temperature (°C), peak time (when peak viscosity occurred; min), peak viscosity (maximum hot paste viscosity), breakdown (peak viscosity minus holding strength or minimum hot paste viscosity), setback (final viscosity minus holding strength or minimum hot paste viscosity), and final viscosity (end of the test after cooling to 50 °C and holding at this temperature) were determined. All the viscosity parameters are expressed in mPa·s.

2.8. Viscoelastic Properties of Starch/Flour–Hydrocolloid Blends

Immediately after pasting, the obtained starch–flour and starch/flour–hydrocolloid gels were subjected to small deformation rheological measurements using an MCR 92 rotational rheometer (Anton Paar GmbH, Inc., Graz, Austria) equipped with a Peltier-temperature-controlled system, as previously suggested by [18]. The test was performed at 50 °C using a 60 mm serrated plate–plate geometry with a 2 mm gap between plates. After lowering the upper plate, the excess sample was trimmed off and a thin layer of silicon oil was used to cover the exposed surface to prevent moisture losses during the test. Prior to starting the test, the experimental gels were rested for 5 min between plates to allow sample relaxation. The frequency sweep test was carried out over the range 0.1–10 Hz at 50 °C using a target strain of 0.01%, which fell within the linear viscoelastic region previously determined by running a strain sweep test at a constant frequency of 10 Hz and with a strain that varied over the range 0.001–100. Values of storage modulus (G'), loss modulus (G''), and loss tangent (tan δ) were recorded at a frequency of 1 Hz.

2.9. Statistical Analysis

The experimental data obtained from viscometric and rheological measurements were analyzed using one-way analysis of variance (ANOVA). Fisher's least significant differences (LSD) test was applied to determine the difference between each pair of means with 95% confidence ($p < 0.05$). Statistical analysis of the results was performed using Statistica v10.0 software (StatSoft, Inc., Tulsa, OK, USA).

3. Results and Discussion

3.1. Structural and Viscosity Properties of the Single Hydrocolloids

Guar, tara, and methylcellulose powders were analyzed by a wide-angle X-ray diffraction technique in order to evaluate the effect of the mechanical treatment on the structural evolution of the three hydrocolloids (Figure 1).

The XRD patterns of the un-milled GG (UN-GG) and milled GG (M-GG) powders are presented in Figure 1A. Both guar systems exhibited a typical XRD profile with very low crystallinity (2θ = 17.02 and 20.11°) with respect to the amorphous counterpart (broad peak centered at 18.1° in 2θ). After the ball milling, a pronounced reduction in crystallinity (see peak at 17.02°), which could be ascribable to the partial hydrogen bonds breaking the guar structure, is observed [19,20].

Figure 1B depicts the XRD pattern of the TG before and after milling. The two broad peaks centered at 18.6 and 39.8° in 2θ are typical of an amorphous pattern. The effect of milling on these powders was not comparable with the previous system: the analysis of the peak shape and broadening revealed an increase of the amorphous component of 2%–4% for the milled sample. For this reason, a clear effect of the mechanical treatment cannot be shown by X-ray diffraction for the TG sample. FT-IR analysis, not shown here, seemed to confirm the experimental evidence achieved by X-ray diffraction.

XDR patterns for MC samples are shown in Figure 1C. The observed diffraction peaks for both materials can be attributed to the crystalline scattering and the diffuse background of the disordered regions. The diffractogram corresponding to the un-milled MC (UN-MC) sample showed maximum diffraction peaks at the 8.1° and 20.2° 2θ angles, while 8.1° and 19.7° 2θ angles were recorded for the ball milled MC (M-MC). Comparing the XRD patterns, it is possible to observe a shift versus a lower 2θ angle of the main peak for the ball-milled system. This indicates an increase of the inter-planar distance compared to the original MC, due to the generation of disorder when the MC is ball milled. The maximum, around 20°, present in the MC sample, is commonly known as van der Waals halo, which appears for many polymers and corresponds to the polymeric chain packing due to van der

Waals forces [21]. Also, the maximum around 8°, which is known as the halo of low van der Waals, occurs for some amorphous polymers due to the existence of regions with aggregates of segments of parallel chains. The peak broadening (b–violet curve) is a direct consequence of the mechanical treatment which induces a further decrease of the crystallinity of the MC powders (35%). In the case of the native MC, this maximum is much more defined, conferring this sample a more semicrystalline character (46%).

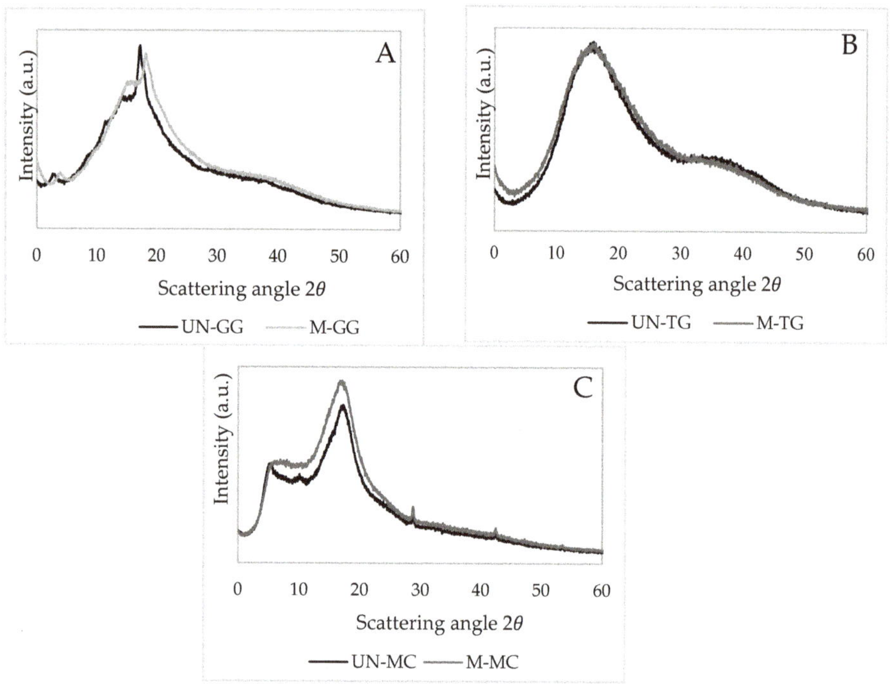

Figure 1. X-ray diffraction patterns of un-milled (continuous dark line) and milled (continuous grey line) hydrocolloids: guar gum (**A**), tara gum (**B**), and methylcellulose (**C**). UN: un-milled; M: milled; GG: guar gum; TG: tara gum; MC: methylcellulose.

The gelling behavior of the investigated hydrocolloids, both before and after milling, were then analyzed as a function of the time to determine potential changes in the viscosity properties of the hydrocolloid aqueous solutions.

The difference in viscosity due to the ball milling treatment is clearly visible in Figure 2. In this case, the effect of ball milling on the guar gum resulted in a diminished viscosity due to the fact that the crystalline domains decreased in length and therefore the chains flowed better (Figure 2A). On the contrary, the greater affinity due to the overexposure of OH groups of the hydrocolloid chains turned out to be the dominant effect in tara gum aqueous solutions, which showed increased values of viscosity (Figure 2B).

Methylcellulose appeared to have an intermediate behavior between the two effects, as if they were subjected to compensation. In fact, there was no substantial difference between the UN-MC and the M-MC. The RVA analysis graph (Figure 2C) showed an important characteristic of methylcellulose: its lower critical solution temperature (LCST). LCST is a change in the MC microstructure, passing from a hydrophilic to a hydrophobic state as a consequence of a change of temperature [22]. The behavior of this hydrocolloid is not monotonous when the temperature varies; the viscosity of an MC solution slightly decreases below a critical temperature value over which the viscosity regime increases [23]. In the RVA graph, the change in viscosity was between 40 and 50 °C.

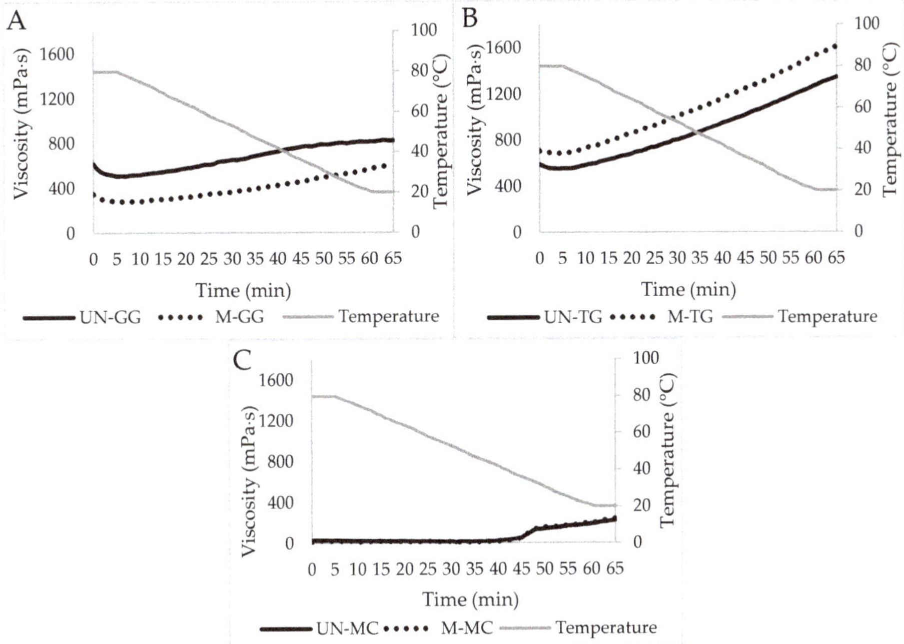

Figure 2. Gelling/thickening profiles of the three un-milled (continuous line) and milled (discontinuous line) hydrocolloids: guar gum (**A**), tara gum (**B**), and methylcellulose (**C**). UN: un-milled; M: milled; GG: guar gum; TG: tara gum; MC: methylcellulose.

3.2. Properties of Binary Hydrocolloid–Starch Blends

The viscometric properties of a corn starch-rice flour system in the absence and presence of milled and un-milled hydrocolloids are reported in Table 2. The effects of both pure hydrocolloids and the ball milling treatment are explored first, prior to analyzing the influence of different combinations of pairs of hydrocolloids on the pasting profiles of the resulting starch–hydrocolloid blends.

Among the tested samples, the starch–flour system alone exhibited lower values of peak viscosity, breakdown, setback, and final viscosity than those observed when the single hydrocolloids were added to the starch paste. Only the MC, irrespective of whether it had been treated or not, showed values of setback and final viscosity during the gelling/cooling stage similar to those of the individual starch (Table 2). A possible explanation for such an increase in viscosity could be that starch–hydrocolloid dispersions are complex biphasic systems composed of a suspension of swollen granules dispersed in a continuous medium in which amylose, low-molecular-weight amylopectin molecules and the added hydrocolloids are located [24,25]. As gelatinization and pasting proceed, the starch granules continue to swell and, in turn, to absorb and bind more water, leading to a reduction of the volume occupied by the hydrocolloids that, being more concentrated, cause a rise in the viscosity of the entire system [24]. Furthermore, the thickening ability of hydrocolloids as well as interactions between hydrocolloids and swollen starch granules and/or leached amylose may also be involved as other concurrent factors [26]. However, while the addition of both milled and un-milled hydrocolloids was significant in promoting the viscosity levels, each hydrocolloid affected the pasting properties of the starch system in a different way (Figure 3). In fact, during gelatinization and pasting, the viscosity profiles of the starch/GG mixtures were higher than those shown by the other galactomannan TG regarding peak viscosity and breakdown, whereas intermediate viscosity values were observed with the addition of the MC, in which the values of peak viscosity were significantly higher than those of the TG mixtures, but the values of breakdown were significantly lower than those of guar gum blends. Discussing the results more in depth, it can be observed that within each tested hydrocolloid, the effect of the ball milling treatment was found to be significant only in the case of GG, in which the mechanochemical treatment seemed to have some detrimental effect on peak viscosity (Table 2). However, the UN-GG was the hydrocolloid that most strongly affected both peak viscosity and breakdown of the starch–flour mixture. This result could be explained by assuming that, during pasting, this hydrocolloid is not able to cover or interact with starch granules, but persists in the macromolecular medium which forms a sheet structure [27]; in this way, it does not hinder the swelling of the starch granules, which are free to swell, leading to a great increase in peak viscosity. After that, the more fragile swollen granules, becoming less resistant to the mechanical shearing, lose their integrity, leading to a severe drop (the so-called breakdown) in the viscosity of the system that becomes, in turn, less stable [25]. The M-GG, on the contrary, when added to the corn starch-rice flour mixture behaved in a different way, resulting in a lower and flatter peak viscosity. To better understand this different behavior, it must be borne in mind that, due to the reduction of the crystalline domains caused by the ball milling treatment, the M-GG exhibited more fluidity than the UN-GG. Therefore, it can be hypothesized that this decrease in peak viscosity (in line with the results obtained in the gelling capacity analysis) may be due to the smooth flow that the M-GG molecules exhibit in the continuous phase without interacting with the leached amylose and amylopectin [28]. Similar conclusions were reported by [28] to explain the decrease in the peak viscosity observed when low-molecular-weight GG was added to a corn starch–rice flour system.

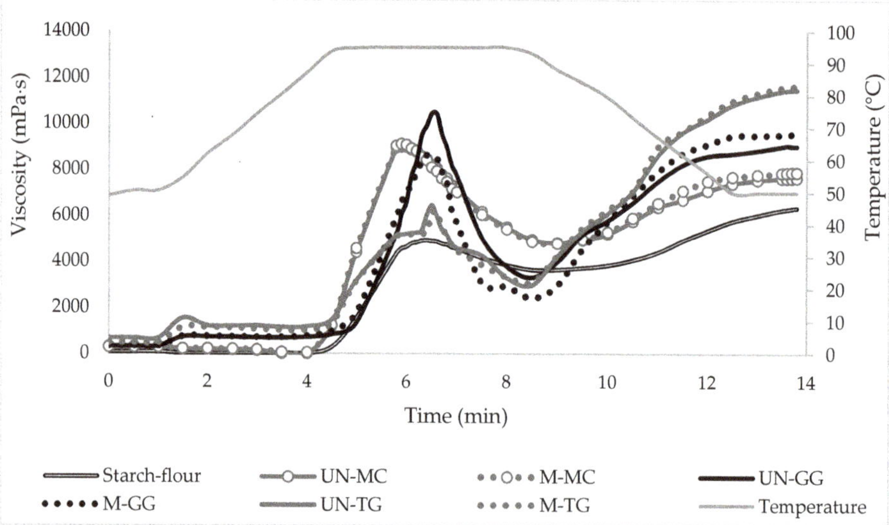

Figure 3. Viscometric profiles of the corn starch/rice flour system in the absence and presence of single milled and un-milled hydrocolloids. UN: un-milled; M: milled; GG: guar gum; TG: tara gum; MC: methylcellulose.

Table 2. Viscometric parameters of corn starch–rice flour systems formulated with one or two un-milled and milled hydrocolloids.

Samples	Parameters																													
	Peak Viscosity (mPa·s)				Breakdown (mPa·s)				Final Viscosity (mPa·s)				Setback (mPa·s)				Peak Time min				Pasting Temperature °C			
Corn starch–rice flour	4917	±	192	a	1300	±	54	a	6360	±	328	a	2743	±	82	a	10.53	±	0.0	ef	77.08	±	0.60	ns
Single Hydrocolloid + Starch–flour																								
UN-GG	10,431	±	430	ef	7168	±	130	cde	9026	±	86	b	5763	±	385	bc	10.70	±	0.0	f	78.23	±	2.30	ns
M-GG	8681	±	730	c	6337	±	1080	c	9529	±	645	bc	7185	±	296	cd	10.57	±	0.0	f	76.30	±	2.90	ns
UN-TG	6732	±	730	b	3854	±	203	b	11,493	±	1520	c	8615	±	1201	d	10.60	±	0.0	f	76.68	±	0.00	ns
M-TG	6436	±	613	b	3508	±	185	b	11,651	±	1641	c	8723	±	1213	d	10.70	±	0.0	f	76.60	±	1.00	ns
UN-MC	9031	±	716	cd	4224	±	938	b	7644	±	5	ab	2837	±	216	a	10.00	±	0.2	abc	76.33	±	0.60	ns
M-MC	9115	±	270	cd	4362	±	412	b	7912	±	186	ab	3158	±	328	a	10.07	±	0.1	c	76.65	±	1.20	ns
Pairs of Hydrocolloids + Starch–flour																								
UN-MC + UN-GG	10,428	±	107	ef	6586	±	35	cd	7299	±	370	ab	3457	±	228	a	10.33	±	0.0	de	78.20	±	1.20	ns
UN-MC + M-GG	13,336	±	434	g	10,527	±	533	g	7277	±	231	ab	4468	±	132	ab	10.04	±	0.0	bc	78.58	±	0.60	ns
M-MC + M-GG	12,980	±	129	g	10,934	±	18	g	6480	±	2948	a	4434	±	2801	ab	9.80	±	0.2	a	78.60	±	0.50	ns
UN-MC + UN-TG	9778	±	454	de	7526	±	260	de	8246	±	204	ab	5994	±	9	bc	10.17	±	0.0	cd	78.70	±	1.80	ns
UN-MC + M-TG	10,573	±	134	ef	7981	±	362	e	8619	±	88	ab	6028	±	408	bc	10.10	±	0.0	c	79.45	±	0.60	ns
M-MC + M-TG	10,948	±	202	f	9235	±	418	f	7445	±	148	ab	5731	±	68	bc	9.84	±	0.2	ab	77.83	±	1.60	ns

Mean values ± standard deviation. Within columns, values (mean of two replicates) followed by the same letter do not differ significantly from each other ($p \leq 0.05$). UN: un-milled; M: milled; GG: guar gum; TG: tara gum; MC: methylcellulose.

As in the case of the GG, the TG, irrespective of whether it had been treated or not, also increased the peak viscosity, but the extent of this increase was significantly lower than that observed for GG (Table 2). Furthermore, while the TG showed a lower breaking viscosity during heating, it promoted higher stability of the system. The authors of [26], when studying the interactions between wheat starch and both guar and tara gums, reported opposite results to our findings. In fact, these authors observed higher values of peak viscosity using the TG, whereas they found similar values of breakdown and setback in the TG-containing samples compared to the guar gum mixtures. This contrasting behavior, however, may be due to the different type of starch and/or to the different method of sample preparation used. Furthermore, in contrast to what was previously reported for the gelling capacity of the milled TG (M-TG) dispersed in an aqueous environment, the ball milling treatment seemed to have no effect on the pasting properties of the TG-starch blend. Presumably, in a complex system such as starch–hydrocolloid mixtures, the overexposure of the hydroxyl groups previously hypothesized for the TG as a consequence of the ball milling treatment, was counterbalanced by the competition for water with the starch granules.

When hot pastes are cooled to 50 °C, a transition from a viscous liquid to a gel, the so-called setback, takes place, leading to an increase in the viscosity of the system primarily due to the re-association of amylose molecules [29]. During this stage, the values of setback and final viscosity were found to be higher than those of the individual starch system for both GG and TG, suggesting that both galactomannans may have a marked tendency to promote starch retrogradation (Table 2). In the case of the UN-GG, however, the extent of such an increase was significantly lower (Figure 3). The differences in the molecular structure of these two galactomannans, especially in terms of the mannose/galactose ratio (3:1 for TG and 2:1 for GG) [10], could explain the differences in the gelling properties of the resulting starch–hydrocolloids blends. In fact, as reported by [30], the presence of a higher number of galactose side chains, which impart molecular flexibility, may facilitate intermolecular associations between guar gum and starch components, delaying, in turn, starch retrogradation of the resulting blends. Moreover, an increase in the effective concentration of amylose in the continuous phase, which leads to faster amylose gelation, could be considered as one of the main factors involved in promoting starch retrogradation [25,26].

Unlike the galactomannans, the addition of the cellulose derivative MC, irrespective of whether it had been treated or not, affected the pasting properties of the starch–flour system only during the heating stage (Figure 3). In fact, values of setback and final viscosity were found to be similar to those of the starch–flour system alone. As previously seen for both the GG and TG, the MC also increased the peak viscosity but, in this case, the amount of time to reach the peak was shorter (Table 2).

However, considering that the MC has the unique property needed to forming gels at temperatures above 60 °C [9], this early increase in the peak viscosity might be related to the hydrocolloid gelation rather than to an effective ability of the MC to assist granule swelling during heating. On the contrary, as previously observed by [31] in potato starch-MC composites, the ability of the MC to prevent starch retrogradation could be explained due to its water solubility. Presumably, the MC, acting as a water binder, could cause either the reduction of the available water required by amylose for the recrystallization process to occur or the inhibition of the possible cross-links between the hydrocolloid and the amylose molecules.

In summary, it can be said that both a cellulose derivative and galactomannans modified the gelatinization and retrogradation behavior of the corn starch–rice flour system in ways that vary depending on their different molecular structure and functionality. On the contrary, the conformational changes induced by mechanochemical treatment have not proven to be effective in changing the behavior of the tested hydrocolloids, except in the case of guar gum.

3.3. Pasting Properties of Ternary Hydrocolloid–Hydrocolloid–Starch Blends

Considering that, in many food systems, the inclusion of a single hydrocolloid might be not sufficient to develop final products with specific attributes in terms of structure, viscosity, stability, and functionality, the effects of combinations of two hydrocolloids on the viscometric properties of a corn starch–rice flour system were also analyzed. It is common agreement that the blending of two hydrocolloids with different functionality in food systems, such as gelling and thickening agents, can have synergistic effects, conferring enhanced viscosity and improved or induced gelation [32,33]. In this regard, six different combinations of both treated and un-treated hydrocolloids were studied: firstly, the UN-MC was combined with each of the two tested galactomannans either before and after the ball milling treatment and, secondly, the M-MC was combined with each of the milled galactomannans.

As reported in Figure 4, the addition of two hydrocolloids, instead of one, significantly changed the viscometric profile of the starch–flour system during both heating and cooling cycles, not only when compared to the individual starch, but also in comparison with the binary blends (Table 2). Among the tested samples, no significant differences were observed only in the pasting temperature parameter, whereas, as stated above, the decreased times to reach the peak viscosity observed in all the experimental blends prepared with the addition of MC were probably due to the thermogelation properties of this cellulose derivative.

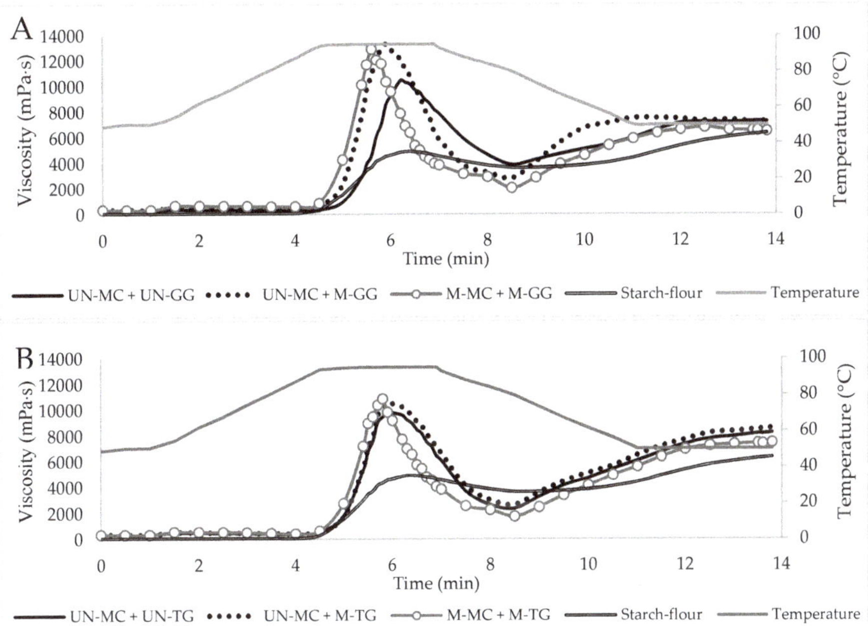

Figure 4. Viscometric profiles of the corn starch-rice flour system in the absence and presence of two milled and un-milled hydrocolloids. MC-GG-starch/flour (**A**) and MC-TG-starch/flour (**B**) ternary mixtures. UN: un-milled; M: milled; GG: guar gum; TG: tara gum; MC: methylcellulose.

Analyzing the results more in depth, it can be observed that the highest peak viscosity and breakdown were obtained in those blends in which the milled guar gum was combined with both the M- and UN-MC (Table 2). These results, being even higher than those that occurred in each hydrocolloid individually, suggested additive interactions between these two gums. A possible explanation for such an increase in viscosity could be that the MC, acting as a gelling agent, is able to build up a network in which the GG, acting as thickening agent, fills up the spaces [32]; in this way, the concentration of the continuous phase increases, causing, in turn, a rise in viscosity of the entire system. Moreover, considering that such an increase was less pronounced when the two hydrocolloids were combined in their native form, it can be hypothesized that the structural changes induced by the ball milling treatments (especially for guar gum) may have facilitated such heterotypic interactions. In addition, during the cooling stage, the values of setback and final viscosity, being lower than those observed for the M-GG/starch mixture, suggested that the addition of both the M- and UN-MC may promote a marked tendency to prevent starch retrogradation of the resulting ternary blends.

As in the case of the M-GG, the TG used in combination with both the M- and UN-MC also increased peak viscosity and breakdown of the starch–flour system, but the effect was less pronounced when the two hydrocolloids were combined in their native form (Figure 4B). During the cooling stage, however, lower values of final viscosity and setback were observed as compared to the binary mixture with added the M-TG, but higher values of setback and similar values of final viscosity were obtained as compared to the binary mixture with both the M- and UN-MC (Table 2). On the basis of these results, it can be hypothesized that, although there have been additive interactions between the two hydrocolloids, the blending of MC and GG seemed to be more effective in modifying the viscometric profile of the corn starch–rice flour system.

3.4. Viscoelastic Properties of Hydrocolloids Starch Gels

The viscoelastic parameters of the corn starch/rice flour system in the absence and presence of milled and un-milled hydrocolloids were measured at 50 °C immediately after pasting.

In all the tested samples, the storage modulus (G′), the elastic component of the material, was higher than the loss modulus (G″), the viscous part of the material, indicating that all the starch pastes had a solid, elastic-like behavior typical of a biopolymer gel (Table 1). However, for all the tested mixtures, values of both dynamic moduli did not differ significantly from those of the starch alone, except for the UN-GG/starch and M-TG/starch gels, in the case of the binary mixtures, and the M-MC/M-TG/starch gels, in the case of the ternary mixtures. Such increases in the viscoelastic properties observed in both mixtures prepared with the M-TG, as well as in the binary mixture prepared with the UN-GG, may be attributed to an increase in the viscoelastic properties of the individual gums. However, the mechanochemical treatment seemed to affect the galactomannans behavior in opposite ways. In fact, in the case of GG, the elastic properties of the starch-hydrocolloid system were found to be higher than those of the individual starch only when the gum was added in its pure form, suggesting that the mechanochemical treatment might have lowered the tendency of the GG to form gels, probably as a result of increased synergistic interactions. On the contrary, in the blends obtained by adding the UN-TG, considering that no differences in the values of both dynamic moduli (apart from a slight increase in the binary mixture) were observed in comparison to the individual starch (Table 3), it can be assumed that the ball milling treatment may have favorably affected the hydrocolloid behavior in this complex system.

Moreover, analyzing the results more in depth, it can be observed that the addition of the M-TG into the starch–flour system led to an increase in G″ values more pronounced than that observed in G′, indicating that this galactomannan was more effective in increasing the viscous properties of the system rather than the elastic properties (Table 3). A possible explanation for these results could be related to the phase separation between starch components and galactomannan that takes place in the dispersed phase as a result of thermodynamic incompatibility between these two chemically dissimilar polymers [24]. In this way, intramolecular interactions may be energetically promoted in comparison with intermolecular associations, modifying the viscoelastic properties of the resulting starch–hydrocolloid gels. Similar conclusions were reported by [34] to explain the increase in the viscous properties of a rice starch–flour system when increasing concentration levels of the TG (0.2%–0.6%) are added to it. In addition, although no significant differences were observed in the tan δ values of all the hydrocolloid-starch pastes with respect to the individual starch, the highest values of the ratio G″/G′ shown by both the M-TG starch mixtures seemed to further confirm the results described above.

Table 3. Viscoelastic properties of corn starch–rice flour systems formulated with one or two un-milled and milled hydrocolloids.

Samples	Storage Modulus G′ (Pa)			Loss Modulus G″ (Pa)			Loss Tangent tan δ			
Corn starch–rice flour	1449	±	191	abc	147	±	0	a	0.103	±	0.013	ns
Single Hydrocolloid + Starch–flour												
UN-GG	2024	±	320	d	234	±	12	ab	0.117	±	0.012	ns
M-GG	1222	±	227	a	151	±	31	a	0.124	±	0.002	ns
UN-TG	1785	±	105	cd	315	±	66	ab	0.176	±	0.026	ns
M-TG	2887	±	173	e	635	±	335	c	0.217	±	0.103	ns
UN-MC	1254	±	161	ab	212	±	20	ab	0.170	±	0.005	ns
M-MC	1287	±	230	ab	239	±	66	ab	0.185	±	0.018	ns
Pairs of Hydrocolloids + Starch–flour												
UN-MC + UN-GG	1594	±	101	abcd	245	±	50	ab	0.153	±	0.021	ns
UN-MC + M-GG	1705	±	201	bcd	252	±	3	ab	0.149	±	0.016	ns
M-MC + M-GG	1816	±	34	cd	284	±	37	ab	0.157	±	0.018	ns
UN-MC + UN-TG	1379	±	14	abc	237	±	6	ab	0.172	±	0.002	ns
UN-MC + M-TG	1751	±	523	cd	312	±	113	ab	0.177	±	0.012	ns
M-MC + M-TG	1942	±	26	d	399	±	110	b	0.206	±	0.059	ns

Mean values ± standard deviation. Within columns, values (mean of two replicates) followed by the same letter do not differ significantly from each other ($p \leq 0.05$). UN: un-milled; M: milled; GG: guar gum; TG: tara gum; MC: methylcellulose.

4. Conclusions

On the basis of the data obtained by XRD experiments and gelling tests, it could be said that the ball milling treatment affected the structure of the tested hydrocolloids and, in turn, the viscosity of their aqueous solutions in different ways. On the one hand, in the case of the GG, the reduction of the crystalline domains and the consequent increase in the fluidity of the system induced by the mechanochemical treatment seemed to be the main causes of the decrease in viscosity observed when the M-GG aqueous solutions were compared with those prepared with the UN-GG; on the other hand, in the case of the TG, the larger exposure of the hydroxyl groups seemed to justify the increase in viscosity observed in the M-TG aqueous solutions when compared to those prepared with the UN-TG, even if it was not possible to confirm this effect with XRD and IR analysis. On the contrary, the MC seemed to have an intermediate behavior between the two effects, leading to aqueous solutions with the same viscosity.

When each hydrocolloid was individually combined with a starch–flour system, these same trends were observed in the viscometric profiles of the starch–hydrocolloid systems, with the exception of starch-TG blends in which no differences between blends containing the M- and the UN-forms were observed. However, considering that the conformational changes induced by the mechanochemical treatment seemed to be effective only in the case of the GG, it could be assumed that the different pasting properties observed in the binary starch–hydrocolloid mixtures in comparison to the starch alone might be due to the type of hydrocolloid added rather than to structural changes induced by the treatment. On the contrary, the ball milling treatment seemed to be more effective when two combined hydrocolloids, especially when they were both in the milled form, were simultaneously added to the starch–flour system. On the contrary, the dynamic moduli of the cornstarch–rice flour gels were significantly affected by the addition of the hydrocolloids only in the case of the UM-GG/starch, M-TG/starch, and M-MC/M-TG/starch blends, indicating that the ball milling treatment was involved in promoting phase separation between starch and M-TG molecules in the dispersed phase to a greater extent in comparison to the guar systems.

This study represents a preliminary attempt to understand how the effects induced by the ball milling treatment on different types of hydrocolloids may change their functionality not only when they are used alone, but also when they are included in a complex system, such as a starch–flour system. Further experiments are needed to analyze more in depth which intra- and inter-molecular interactions, occurring between starch and treated hydrocolloids and within the same hydrocolloid molecules, may be involved in changing the paste and gel behavior of starch–hydrocolloid systems.

Author Contributions: Conceptualization, L.N., P.C., and C.F.; methodology, L.N., P.C., S.G., A.P., and C.F.; formal analysis, L.N., P.C., and V.F.; data curation, L.N., P.C., S.G., V.F., A.P., and C.F.; writing—original draft preparation, L.N., P.C., S.G., V.F., A.P., and C.F.; writing—review and editing, L.N., P.C., S.G., V.F., A.P., and C.F.; supervision, C.P., A.P., and C.F.; project administration, C.F.; funding acquisition, C.F. All authors have read and agreed to the published version of the manuscript.

Funding: This research was funded by Sardegna Ricerche (Capitale Umano ad Alta Qualificazione", 2015—L.R. n°7, 7 agosto 2007). Project title: "Studio di nuovi ingredienti per il miglioramento delle proprietà tecnologiche, sensoriali e nutrizionali di prodotti da forno gluten-free". Valeria Farina and Dr. Luca Nuvoli are supported by a PhD program in a collaborative scheme between the University of Sassari and Cagliari of Italy, which is endorsed by the Autonomous Regional Administration of Sardinia (RAS).

Acknowledgments: The authors acknowledge the support of CeSAR UNISS, "Centro Servizi di Ateneo per la Ricerca" of the University of Sassari, for making the XRD apparatus (SmartLab Rigaku) available to carry out material characterization.

Conflicts of Interest: The authors declare no conflict of interest.

References

1. Biliaderis, C.G.; Arvanitoyannis, I.; Izydorczyk, M.S.; Prokopowich, D.J. Effect of hydrocolloids on gelatinization and structure formation in concentrated waxy maize and wheat starch gels. *Starch/Staerke* **1997**, *49*, 278–283. [CrossRef]
2. Copeland, L.; Blazek, J.; Salman, H.; Tang, M.C. Form and functionality of starch. *Food Hydrocoll.* **2009**, *23*, 1527–1534. [CrossRef]
3. Rosell, C.M.; Yokoyama, W.; Shoemaker, C. Rheology of different hydrocolloids-rice starch blends. Effect of successive heating-cooling cycles. *Carbohydr. Polym.* **2011**, *84*, 373–382. [CrossRef]
4. Gao, Z.; Fang, Y.; Cao, Y.; Liao, H.; Nishinari, K.; Phillips, G.O. Hydrocolloid-food component interactions. *Food Hydrocoll.* **2017**, *68*, 149–156. [CrossRef]
5. Burey, P.; Bhandari, B.R.; Howes, T.; Gidley, M.J. Hydrocolloid gel particles: Formation, characterization, and application. *Crit. Rev. Food Sci. Nutr.* **2008**, *48*, 361–377. [CrossRef]
6. Liu, X.; Mu, T.; Sun, H.; Zhang, M.; Chen, J.; Fauconnier, M.L. Influence of different hydrocolloids on dough thermo-mechanical properties and in vitro starch digestibility of gluten-free steamed bread based on potato flour. *Food Chem.* **2018**, *239*, 1064–1074. [CrossRef]
7. Conte, P.; Fadda, C.; Drabińska, N.; Krupa-Kozak, U. Technological and nutritional challenges, and novelty in gluten-free breadmaking: A review. *Pol. J. Food Nutr. Sci.* **2019**, *69*, 5–21. [CrossRef]
8. Dickinson, E. Hydrocolloids as emulsifiers and emulsion stabilizers. *Food Hydrocoll.* **2009**, *23*, 1473–1482. [CrossRef]
9. Nasatto, P.L.; Pignon, F.; Silveira, J.L.M.; Duarte, M.E.R.; Noseda, M.D.; Rinaudo, M. Methylcellulose, a cellulose derivative with original physical properties and extended applications. *Polymers* **2015**, *7*, 777–803. [CrossRef]
10. Prajapati, V.D.; Jani, G.K.; Moradiya, N.G.; Randeria, N.P.; Nagar, B.J.; Naikwadi, N.N.; Variya, B.C. Galactomannan: A versatile biodegradable seed polysaccharide. *Int. J. Biol. Macromol.* **2013**, *60*, 83–92. [CrossRef]
11. Song, J.Y.; Kwon, J.Y.; Choi, J.; Kim, Y.C.; Shin, M. Pasting properties of non-waxy rice starch–hydrocolloid mixtures. *Starch/Staerke* **2006**, *58*, 223–230. [CrossRef]
12. Rojas, J.A.; Rosell, C.M.; Benedito De Barber, C. Pasting properties of different wheat flour-hydrocolloid systems. *Food Hydrocoll.* **1999**, *13*, 27–33. [CrossRef]
13. Abbaszadeh, A.; Macnaughtan, W.; Foster, T.J. The effect of ball milling and rehydration on powdered mixtures of hydrocolloids. *Carbohydr. Polym.* **2014**, *102*, 978–985. [CrossRef]
14. Lu, X.; Zhang, H.; Li, Y.; Huang, Q. Fabrication of milled cellulose particles-stabilized Pickering emulsions. *Food Hydrocoll.* **2018**, *77*, 427–435. [CrossRef]

15. Tian, Y.; Zhang, K.; Zhou, M.; Wei, Y.J.; Cheng, F.; Lin, Y.; Zhu, P.X. High-Performance Starch Films Reinforced With Microcrystalline Cellulose Made From Eucalyptus Pulp via Ball Milling and Mercerization. *Starch/Staerke* **2019**, *71*, 1800218. [CrossRef]

16. Ramadhan, K.; Foster, T.J. Effects of ball milling on the structural, thermal, and rheological properties of oat bran protein flour. *J. Food Eng.* **2018**, *229*, 50–56. [CrossRef]

17. Avolio, R.; Bonadies, I.; Capitani, D.; Errico, M.E.; Gentile, G.; Avella, M. A multitechnique approach to assess the effect of ball milling on cellulose. *Carbohydr. Polym.* **2012**, *87*, 265–273. [CrossRef]

18. Martínez, M.M.; Macias, A.K.; Belorio, M.L.; Gómez, M. Influence of marine hydrocolloids on extruded and native wheat flour pastes and gels. *Food Hydrocoll.* **2015**, *43*, 172–179. [CrossRef]

19. Gong, H.; Liu, M.; Chen, J.; Han, F.; Gao, C.; Zhang, B. Synthesis and characterization of carboxymethyl guar gum and rheological properties of its solutions. *Carbohydr. Polym.* **2012**, *88*, 1015–1022. [CrossRef]

20. Mudgil, D.; Barak, S.; Khatkar, B.S. X-ray diffraction, IR spectroscopy and thermal characterization of partially hydrolyzed guar gum. *Int. J. Biol. Macromol.* **2012**, *50*, 1035–1039. [CrossRef]

21. Oliveira, R.L.; Vieira, J.G.; Barud, H.S.; Assunção, R.M.N.; Filho, G.R.; Ribeiro, S.J.L.; Messadeqq, Y. Synthesis and characterization of methylcellulose produced from bacterial cellulose under heterogeneous condition. *J. Braz. Chem. Soc.* **2015**, *26*, 1861–1870. [CrossRef]

22. Mariani, A.; Nuvoli, L.; Sanna, D.; Alzari, V.; Nuvoli, D.; Rassu, M.; Malucelli, G. Semi-interpenetrating polymer networks based on crosslinked poly (N-isopropyl acrylamide) and methylcellulose prepared by frontal polymerization. *J. Polym. Sci. Part A Polym. Chem.* **2018**, *56*, 437–443. [CrossRef]

23. Hirrien, M.; Chevillard, C.; Desbrières, J.; Axelos, M.A.V.; Rinaudo, M. Thermogelation of methylcelluloses: New evidence for understanding the gelation mechanism. *Polymer (Guildf)* **1998**, *39*, 6251–6259. [CrossRef]

24. Alloncle, M.; Doublier, J.L. Viscoelastic properties of maize starch/hydrocolloid pastes and gels. *Top. Catal.* **1991**, *5*, 455–467. [CrossRef]

25. Bemiller, J.N. Pasting, paste, and gel properties of starch–hydrocolloid combinations. *Carbohydr. Polym.* **2011**, *86*, 386–423. [CrossRef]

26. Funami, T.; Kataoka, Y.; Omoto, T.; Goto, Y.; Asai, I.; Nishinari, K. Effects of non-ionic polysaccharides on the gelatinization and retrogradation behavior of wheat starch. *Food Hydrocoll.* **2005**, *19*, 1–13. [CrossRef]

27. Chaisawang, M.; Suphantharika, M. Effects of guar gum and xanthan gum additions on physical and rheological properties of cationic tapioca starch. *Carbohydr. Polym.* **2005**, *61*, 288–295. [CrossRef]

28. Funami, T.; Kataoka, Y.; Omoto, T.; Goto, Y.; Asai, I.; Nishinari, K. Food hydrocolloids control the gelatinization and retrogradation behavior of starch. 2a. Functions of guar gums with different molecular weights on the gelatinization behavior of corn starch. *Food Hydrocoll.* **2005**, *19*, 15–24. [CrossRef]

29. Collar, C.; Conte, P.; Fadda, C.; Piga, A. Gluten-free dough-making of specialty breads: Significance of blended starches, flours and additives on dough behaviour. *Food Sci. Technol. Int.* **2015**, *21*, 523–536. [CrossRef]

30. Funami, T. Functions of Food Polysaccharides to Control the Gelatinization and Retrogradation Behaviors of Starch in an Aqueous System in Relation to the Macromolecular Characteristics of Food Polysaccharides. *Food Sci. Technol. Res.* **2010**, *15*, 557–568. [CrossRef]

31. Kohyama, K.; NIshinari, K. Cellulose Derivatives Effects on Gelatinization and Retrogradation of Sweet Potato Starch. *J. Food Sci.* **1992**, *57*, 128–131. [CrossRef]

32. Hoefler, A.C. *Hydrocolloids*; Eagan Press Handbook Series, St.: Paul, MN, USA, 2004.

33. Saha, D.; Bhattacharya, S. Hydrocolloids as thickening and gelling agents in food: A critical review. *J. Food Sci. Technol.* **2010**, *47*, 587–597. [CrossRef] [PubMed]

34. Lee, H.Y.; Jo, W.; Yoo, B. Rheological and microstructural characteristics of rice starch–tara gum mixtures. *Int. J. Food Prop.* **2017**, *20*, 1879–1889. [CrossRef]

Article

Dough Rheological Behavior and Microstructure Characterization of Composite Dough with Wheat and Tomato Seed Flours

Silvia Mironeasa and Georgiana Gabriela Codină *

Faculty of Food Engineering, Stefan cel Mare University of Suceava, 720229 Suceava, Romania; silviam@fia.usv.ro
* Correspondence: codina@fia.usv.ro; Tel.: +40-745460727

Received: 27 October 2019; Accepted: 25 November 2019; Published: 30 November 2019

Abstract: The rheological and microstructural aspects of the dough samples prepared from wheat flour and different levels of tomato seed flour (TSF) were investigated by rheology methods through the Mixolab device, dynamic rheology and epifluorescence light microscopy (EFLM). The Mixolab results indicated that replacing wheat flour with TSF increased dough development time, stability, and viscosity during the initial heating-cooling cycle and decreased alpha amylase activity. The dynamic rheological data showed that the storage modulus G' and loss modulus G'' increased with the level of TSF addition. Creep-recovery tests of the samples indicated that dough elastic recovery was in a high percentage after stress removal for all the samples in which TSF was incorporated in wheat flour. Using EFLM all the samples seemed homogeneous showing a compact dough matrix structure. The parameters measured with Mixolab during mixing were in agreement with the dynamic rheological data and in accordance with the EFLM structure images. These results are useful for bakery producers in order to develop new products in which tomato seed flour may be incorporated especially for wheat flours of a good quality for bread making and high wet gluten content. The addition of TSF may have a strength effect on the dough system and will increase the nutritional value of the bakery products.

Keywords: tomato seed flour; wheat flour; dough rheology; microstructure

---

1. Introduction

Tomato (*Lycopersicon esculentum*) is one of the most consumed vegetables, as raw (fresh), cooked, in food preparations and as processed products such as tomato juice, puree, paste, ketchup, sauce and canned tomato. Due to its nutritional and bioactive components, the foods products in which tomato is incorporated represents a valuable source of minerals, vitamins and antioxidants, namely lycopene, which varies according to the tomato variety, maturity and agro-environmental factors during growth [1,2]. Post-harvest storage conditions [3,4] and processing technology [5] also influence the tomatoes and tomato-based food products. Dietary intake of tomatoes and processed tomato products has been associated with a decrease in susceptibility to chronic diseases such as various type of cancers and cardiovascular diseases [6]. The medical or health benefits of tomato intake are attributed to the natural antioxidant compounds present in tomato such as ascorbic acid, carotenoids (mainly lycopene) and phenolic compounds which play a crucial role in the health protection mechanisms by scavenging free radicals [7].

Processing of tomatoes leads to a high amount of by-products, about 3–7% of the tomato weight [8,9] that generates serious environmental problems for the industry concerned due to the disposal of the organic material. Tomato pomace, the major part of the by-products, includes mainly seeds and peels in various proportions and a small amount of pulp [10]. Some amounts of by-products are used in animal feed or as soil fertilizers [11]. However, today there is an increasing trend of using this by-product as a source of functional components in different products knowing that the tomato pomace is a rich source of valuable compounds which can be recovered and used in food, pharmaceutical and cosmetic industries as natural ingredients [12,13]. In the chemical composition of tomato pomace high level of total dietary fibers, proteins, fats and medium amounts of ash are found [14].

The peel and seed by-products of tomato pomace represents approximately 20–50 g·kg^{-1} of the initial weight of tomatoes and can be used individually or combined [14]. Regarding the combined utilization, the literature reports that, due to the significant amounts of bioactive phytochemicals from tomato pomace, it can be used as natural antioxidants for the formulation of functional foods, or as additives in food systems to extend their shelf-life [15–17]. Other studies showed that tomato pomace can be considered a good source of some macroelements, such as potassium, manganese and calcium and microelements, i.e., copper and zinc, which are cofactors of the antioxidant enzymes [15]. As individual part of by products, many studies [16–19] highlighted the possibility of using the tomato peels as a source of carotenoids and antioxidants to enrich various foods products or to produce new functional ones. The other part of tomato by-product, tomato seeds, account for approximately 10% of the fruit and 60% of the total by-product and contain a high amount for protein, fat and mineral elements such as potassium, calcium, iron, manganese, zinc and copper [14,20]. Tomato seed proteins present adequate properties to form a good emulsion in a food system [21]. An improved protein quality for bread supplemented with 10% seed meal was reported by Sogi et al. [22]. According to Kramer and Kwee [23] the nutritive value of tomato seed protein is lower than of casein but equivalent or higher than of other plant proteins. The net protein retention (NPR) of whole tomato seed meal, defatted tomato seed meal and tomato seed protein concentrate was reported as 2.65%, 2.52% and 2.51%, respectively which were lower but comparable with those obtained for casein, 2.91% (dry basis) [24].

High levels of essential amino-acids present in tomato seeds showed that this by-product presents high quality proteins, with a high amount of lysine (3.4–5.9%) [25–27]. The amount of lysine in tomato seed is approximately 13% higher than the amount found in the soy protein [28]. Therefore, the tomato seed may be recommended to fortify various low-lysine food products that are deficient in this amino acid. One of these products are the bakery ones of which the base raw material is wheat flour which contains low amounts of lysine. In addition, compared to other seed sources, no anti-nutritional constituents have been reported in tomato seeds [29], a fact that make them a better source of proteins compared to other non-conventional sources.

The use of tomato seeds in order to improve the quality of food products were less studied. These studies were focused especially on the use of the dried pomace [30,31] or tomato peels in food products [32,33]. However, the use of tomato seed in bread making has been previously reported in a few studies [34–36] which were especially referring to the tomato seed addition effect on bread quality. To our knowledge, the information on the effect of partial substitution of wheat flour with whole tomato seed flour on dough rheological properties are missing. No studies have been made on the effect of wheat flour substitution with tomato seed flour at the levels of 5%, 10%, 15% and 20% on dough rheological properties and its microstructure. The use of this kind of methodology, which investigates both empirical and fundamental rheological properties on this subject, is not frequent. Also, the investigation of dough microstructure through a modern device namely epifluorescence light microscopy (EFLM) helped us to better understand the wheat flour dough behavior with different levels of tomato seed flour (TSF) addition during the bread making technological process. The objective of this study was: (1) to evaluate the physicochemical characteristics for the wheat–tomato seed composite flours; (2) to investigate the changes that occur in the dough formed by the wheat–tomato seed composite flours during mixing as well as the quality of starch and protein from the dough system by using the Mixolab device; (3) to analyze the effect of tomato seed flour addition (TSF) on the fundamental dough rheological properties by using a rheometer on which oscillatory frequency test and creep recovery test was performed; (4) to analyze the effect of TSF on dough microstructure, further analyzed through epifluorescence light microscopy (EFLM).

2. Materials and Methods

2.1. Flour Samples

One commercial refined wheat flour type 650 was used. The wheat flour was provided by S.C. Mopan S.A. Company (Suceava, Romania), a local milling company. Tomato seed flour was collected from tomatoes (*Solanum lycopersicum*) cultivated in Suceava County, Romania. The tomato seeds were separated and cleaned after our own developed procedure from the tomato by-product (peel, pulp and seeds) with water, at the temperature of 24 °C, after which they were dried in a tray dryer (Memmert UF30, City, Schwabach, Germany) at 50 °C until its moisture content reached less than 10% (wet basis). This was to limit the loss of available bioactive compounds which are still available even at high levels of drying temperature, namely 60 °C according to Nour et al. [15] or 70 °C according to Sogi et al. [22]. After cooling, the tomato seed were ground in an electrical mill (Heinner, Navy 150, Guangdong City, China). Sieving was made on sieves using a vibratory shaker (Retsch Vibratory Sieve Shaker AS 200 basic, Haan, Germany) in order to obtain particle sizes lower than 500 μm. The raw materials were analyzed according to the international or Romanian standard methods. The moisture content was determined according to International Association for Cereal Chemistry (ICC) method 110/1, ash content according to ICC 104/1, protein content according to ICC 105/2, fat content according to ICC 136. The wheat flour mixes (blends) were analyzed also for the falling number value according to ICC 107/1. The gluten deformation and wet gluten content were analyzed according to the Romanian standard SR 90/2007.

2.2. Flour Composites

Flour composites were obtained from the wheat flours which were substituted by tomato seed flour at the levels of 0%, 5%, 10%, 15% and 20%. The wheat–tomato seed composite flours were analyzed for their moisture content according to ICC 110/1, fat content according to ICC 136, protein content according to 105/2, ash content according to ICC 104/1 and falling number according to ICC 107/1. According to previously studies [15] it seems that the incorporation of tomato pomace in wheat flour up to 10% level did not have a negative effect on the acceptability of bread. Also, according to Carlson et al. [34], from the technological point of view, high levels of 20% tomato seed flour addition in wheat flour increased loaf volume of bread. However, higher levels such as 25% tomato pomace addition in wheat flour presented a negative effect on the flavor quality of bakery products as Bhat and Ahsan [30] reported.

2.3. Evaluation of Wheat–Tomato Seed Composite Flours on Mixolab Dough Rheological Properties

The Mixolab device (Chopin, Tripette et Renaud, Paris, France) was used to analyze the mixing and pasting behavior of wheat–tomato seed composite flours according to ICC standard method No.173. The tests were made for each sample in order to achieve the optimum dough consistency of 1.1 Nm. The evaluated parameters from the Mixolab curve were the water absorption capacity (WA), dough development time (DT), dough stability (ST), minimum torque value corresponding to the initial heating (C2), maximum torque value corresponding to the heating stage (C3), torque value corresponding to the stability of hot starch paste (C4), torque value corresponding to the final starch paste viscosity after cooling (C5) and the difference between Mixolab torques C1 and C2 (C1-2), C2 and C3 (C3-2), C3 and C4 (C3-4), C4 and C5 (C5-4).

2.4. Evaluation of Wheat–Tomato Seed Composite Flours on Rheological Properties

Oscillatory and creep and recovery tests were performed at 25 °C using a HAAKE MARS 40 rheometer (Termo-HAAKE, Karlsruhe, Germany) with non-serrated parallel plate geometry with a diameter of 40 mm and a gap width of 2 mm.

The dough samples were prepared at optimum Farinograph water absorption by mixing until full dough development by using a Brabender Farinograph®-E with a 300 g capacity (Brabender OHG, Duisburg, Germany). The composite flour of 14% moisture basis was kept in the Farinograph bowl and during mixing water was added from the burette to give a dough consistency of 500 BU. Then, the sample prepared was placed between plates and rested for 5 min to allow relaxation and temperature stabilization. The mechanical spectra were measured in a range of linear viscoelasticity, at constant stress of 15 Pa, in frequency sweeps from 1 to 20 Hz. The range of linear viscoelasticity was established based on the dependence of storage modulus (G') and loss modulus (G'') on stress in the region 0.01–100 Pa, at constant oscillation frequency of 1 Hz. The experimental data acquired were described by the power model [37,38]:

$$G'(\omega) = K' \cdot \omega^{n'} \tag{1}$$

$$G''(\omega) = K'' \cdot \omega^{n''} \tag{2}$$

where: G' is storage modulus (Pa), G'' is loss modulus (Pa), ω is angular frequency (rad/s), K', K'' (Pa·s$^{n'}$), n', n'' are experimental constants.

Creep-recovery tests were performed in the range of the linear viscoelasticity at constant stress of 50 Pa. The creep stage time was of 60 s, and the recovery stage was 180 s. The obtaining data of strain as a function of time were expressed in the form of compliance using the following equation [39]:

$$J(t) = \gamma(t)/\sigma \tag{3}$$

where J (Pa^{-1}) is compliance, γ is the strain and σ is the constant stress applied during the creep test (Pa^{-1}).

Experimental data from creep and recovery tests were analyzed by means of a compliance rheological parameter and fitted to the parameter Burger model [40,41] using the Equation (4) for the creep phase and Equation (5) for the recovery phase.

$$J(t) = J_{Co} + J_{Cm}\,(1 - \exp(-t/\lambda_C)) + t/\mu_{Co} \tag{4}$$

$$J(t) = J_{max} - J_{Ro} - J_{Rm}(1 - \exp(-t/\lambda_R)) \tag{5}$$

where J_{io} (Pa^{-1}) is the instantaneous compliance, J_{im} (Pa^{-1}) is the retarded elastic compliance or viscoelastic compliance, t (s) is the phase time, λ_i (s) is the retardation time, μ_{Co} (Pa·s) is the zero shear viscosity and J_{max} (Pa^{-1}) is the maximum creep compliance obtained at the end of the creep test. The recovery compliance, J_r (Pa^{-1}), evaluated where dough recovery reached equilibrium, is calculated by the sum of J_{Ro} and J_{Rm}. The relative elastic part of the maximum creep compliance, expressed as percent recovery was determined using Equation (6) [42,43]:

$$Recovery\ (\%) = \frac{J_r}{J_{max}} \cdot 100 \tag{6}$$

where: J_{max} is the maximum creep compliance value in the creep phase for the 60 s, which corresponds to the maximum deformation, and J_r is the compliance value at the end of the recovery phase.

2.5. Microstructure of Flour Composite Dough

Epifluorescence light microscopy (EFLM) was used to characterize the microstructure of wheat flour dough with different levels of tomato seed flour addition. Dough microstructure was analyzed using a Motic AE 31 inverted microscope (Motic, Optic Industrial Group, Xiamen, China) operated by catadioptric objectives LWD PH 203 (N.A. 0.4). A thin portion was cut from the dough sample and dipped in a fixing solution composed of 1% rhodamine B and 0.5% fluorescein (FITC) in 2-methoxyethanol obtained from Sigma-Aldrich, Germany for at least 1 h. Fluorescein and rhodamine B was used as two fluorescent dyes specific for detecting starch and proteins in the dough samples. Fluorescein detects starch and rhodamine B proteins from the dough system. The EFLM images were analyzed using ImageJ (v. 1.45, National Institutes of Health, Bethesda, MD, USA) software according to Peighambardoust et al. [44] and Codină and Mironeasa [45,46].

2.6. Statistical Analysis

The experimental data and fitting parameters of the employed models were statistically evaluated by one-way analysis of variance (ANOVA) followed by the least significant difference (LSD) test at significance level of 0.05 using SPSS software (trial version, IBM, Armonk, NY, USA). The goodness of fit of the models was assessed using the corresponding determination coefficients (R^2).

3. Results and Discussions

3.1. Flour Characteristics

The analytical characteristics of wheat flour samples were as following: 8.00 mm gluten deformation index, 58.50–62.90% water absorption, 14.50% moisture, 33.00% wet gluten and 0.65% ash content. According to the wheat flour analytical data these were of a very good quality for bread making [47]. The wheat flour used in our study presented high FN values (445 s) indicating that they had a low α amylase activity.

The tomato seed flour presented the following chemical characteristics (dry basis): 6.94/100g moisture content, 29.50/100g protein content, 19.50/100g fat content and 3.92/100g ash content. According to the data obtained, the tomato seed flour was rich in fat and proteins, these values being in agreement with those reported by Mechmeche et al. [25] and Del Valle et al. [9] for tomato seeds.

3.2. Wheat–Tomato Seed Composite Flours Physicochemical Characteristics

The physicochemical characteristics of the wheat–tomato seed composite flours are shown in Table 1. As it may be seen in the composite flours the fat, protein and ash increased with the increase of tomato seed flour addition level whereas the moisture decreased. This was expectable due to the fact that tomato seed flour has a higher amount of fat, proteins and ash and a lower moisture value compared to wheat flour. For the samples in which 20% tomato seed flour was incorporated the values of the fat and ash content were tripled and doubled compared to the control sample whereas the protein content values increased by 27.5%.

The Falling Number (FN) values increased at an addition level up to 10% tomato seed after which its values slightly decreased. This fact indicates that when high levels of TSF were added in wheat flour the blend flour slurry viscosity began to decrease probably due to the decreased viscosity of the starch amount from the samples. It is well known that the flour slurry given by the FN is inversely correlated with α amylase activity [48]. At low levels of TSF addition, the flour mixes viscosity increased and at high levels it slightly decreased, leading to different FN values.

Table 1. Physicochemical characteristics for the wheat–tomato seed composite flours.

Sample	Protein (%)	Lipids (%)	Ash (%)	Moisture (%)	Falling Number (s)
Control	12.40 a (0.20)	1.60 a (0.10)	0.65 a (0.01)	14.50 a (0.30)	445.00 a (14.00)
TSF_5	13.25 b (0.19)	2.49 b (0.09)	0.80 a (0.00)	13.83 a (0.29)	467.00 ab (14.00)
TSF_10	14.11 c (0.18)	3.39 c (0.09)	0.97 a (0.00)	13.47 d (0.27)	478.00 b (12.00)
TSF_15	14.96 d (0.17)	4.28 d (0.09)	1.13 a (0.00)	13.11 c (0.25)	445.00 a (15.00)
TSF_20	15.82 e (0.16)	5.18 e (0.08)	1.30 b (0.00)	12.74 b (0.24)	434.00 ac (16.00)

Values in parentheses are standard deviations. [abcde] Values with the same letter are not significantly different according to the least significant difference (LSD) test ($p < 0.05$).

3.3. Influence of Tomato Seed Flour on Mixolab Dough Rheological Properties

The addition of tomato seed flour (TSF) to wheat flour dough decreased the water absorption values (Table 2) which were 6.2% lower for the dough sample with 20% TSF addition compared to the control. This decrease of the wheat–tomato seed composite flours water absorption may be attributed to the decrease of the gluten content in the blends, which needs less hydration as a result of TSF incorporation, which is gluten free flour [49]. Also, lipids from TSF may partially coat the starch granules and gluten proteins, decreasing the water absorption during mixing [50]. In the mixing stage, both Mixolab parameters dough development time (DT) and stability (ST) increased for the TSF addition up to 20% and 10% respectively. Contrary, a decrease of the farinograph parameters, dough development and stability was reported by Majzoobi et al. [51]. The increase of DT and ST was significant ($p < 0.05$) and may be attributed to the TSF capacity of foaming and emulsifying properties [5]. This will favor the foaming and emulsifying activity of proteins from the dough system, leading to a more stable three-dimensional network structure [52]. The TSF contains a high amount of lipids which interact with proteins and starch. These interactions are facilitated by TSF emulsifying capacity which can bridge the lipids to the gluten proteins or to the starch granules helping the dough to become more stable and structured [53]. However, at levels higher than 10% TSF addition to wheat flour the ST started to decrease probably due to the gluten dilution from the dough system. Compared to the control sample, the dough weakening was insignificant since for all the samples with TSF addition the ST values were higher.

Table 2. Water absorption and Mixolab parameters of tomato seed–wheat flour blends.

Sample	WA (%)	ST (min)	DT (min)	C2 (N-m)	C1-2 (N-m)	C3 (N-m)	C3-2 (N-m)	C4 (N-m)	C3-4 (N-m)	C5 (N-m)	C5-4 (N-m)
Control	60.70 a (2.20)	8.53 a (0.01)	3.85 a (0.62)	0.49 a (0.02)	0.60 a (0.02)	1.82 a (0.07)	1.33 a (0.06)	1.55 a (0.09)	0.27 a (0.02)	2.36 a (0.06)	0.81 a (0.03)
TSF_5	59.20 e (0.02)	8.73 e (0.02)	3.98 d (0.03)	0.54 b (0.03)	0.64 b (0.03)	1.73 d (0.06)	1.19 d (0.03)	1.59 d (0.07)	0.14 c (0.02)	2.81 b (0.05)	1.22 b (0.02)
TSF_10	58.50 d (2.20)	9.18 b (0.02)	4.05 c (0.05)	0.55 b (0.03)	0.62 c (0.01)	1.67 c (0.02)	1.12 c (0.01)	1.57 d (0.04)	0.10 b (0.02)	2.78 b (0.05)	1.21 b (0.01)
TSF_15	57.60 c (2.20)	9.08 c (0.01)	4.32 b (0.17)	0.53 c (0.02)	0.64 b (0.02)	1.52 b (0.05)	0.99 b (0.03)	1.35 b (0.03)	0.17 c (0.02)	2.57 c (0.05)	1.22 b (0.02)
TSF_20	56.80 b (2.20)	8.90 d (0.10)	4.55 b (0.01)	0.53 c (0.02)	0.65 b (0.01)	1.50 b (0.03)	0.97 b (0.01)	1.40 c (0.01)	0.10 b (0.02)	2.53 d (0.04)	1.13 c (0.03)

WA, water absorption; Mixolab parameters: ST, stability; DT, development time; C3, C5, maximum consistency during stage 3, stage 5; C2, C4, minimum consistency during stage 2, stage 4; C1-2, difference of the points C1 and C2; C3-2, difference of the points C3 and C2; C3-4, difference of the points C4 and C3; C5-4, difference of the points C5 and C4. Values in parentheses are standard deviations. [a,b,c,d,e] Values with the same letter are not significantly different according to the LSD test ($p < 0.05$).

The Mixolab parameters related to protein weakening (C2 and difference between the points C1 and C2) were higher for the samples in which tomato seed flour was incorporated (Table 2). These results indicated the fact that dough samples with TSF were stronger than the control sample. However at high levels of TSF addition the C2 value slightly decreased compared to the control sample, probably due to the high gluten dilution of the composite flours.

Tomato seed proteins contain high levels of the globulin fraction [54] which during heating might expose hydrophobic groups which can generate some interaction between proteins, leading to their aggregation [55]. Also, during heating TSF proteins are capable of gel formation leading to a more elastic dough with a more compact protein network [56] which can be breakdown less, a fact reflected by the increased C2 values.

The Mixolab parameters (C3, C4, C5 and the difference between the points C2 and C3 (C3-2), C3 and C4 (C3-4), C4 and C5 (C5-4) are significantly related to the starch behavior during the different heating and cooling phases [57]. The Mixolab C3 and C3-2 values are associated with the starch gelatinization process. The decrease of these values with the increase level of TSF addition showed that the gelatinization capacity of the dough samples decreased (Table 2). This may be attributed to lower starch content from the wheat–tomato seeds composite flour samples. As the amount of TSF in wheat flour increased, the non-starch component from the dough system decreased, leading to a decrease of the dough viscosity when temperature increased above the starch gelatinization.

The C4 torque reflects the hot starch stability paste was slightly increased by 2.58% at levels up to 5% TSF (Table 2). The initial increase of C4 was an expectable one since this value is related to the amylolitic activity from the dough system which decreased with the amount of increased TSF addition. This fact leads to an increased dough viscosity and thus to an increase of the C4 value. Also, compared to the control sample, the difference between C3 and C4 value (C3-4) presented lower values for the samples with TSF addition, a fact that reflects a decrease of the starch degradation rate. This, along with the C4 increase, indicates a more stable gel [58]. However, at high levels of TSF addition the C4 value decreased, this value being 9.67% lower than of the control sample when 20% TSF was incorporated in the dough system. These results may be associated to the protein denaturation from the TSF which according to Sarkar et al. [55] are optimum in the temperature range of 80–90 °C. According to previous reports [55] at the temperature of 87 °C the tomato seed protein showed significant denaturation caused by its globular and non-globular protein components. The protein denaturation from TSF will lead to changes in the dough system conformation of which viscosity begins to decrease. Also, by its denaturation the TSF proteins lose their capacity to retain water. This can be retained in a higher amount by the starch granules of which the gelatinization process becomes more complete, a fact that may favor its amylolitic atacability.

For the dough samples in which TSF were added in wheat flour, C5 and the difference between the C5 and C4 peaks (C5-4) presented higher values than for the control (Table 2). This fact indicated that by TSF addition in wheat flour the degree of starch retrogradation increased. However, when high levels of TSF were added, the C5 and C5-4 began to decrease probably due to the lipid content of the TSF which may have interacted with starch and proteins from the dough system, favoring less starch retrogradation [59].

3.4. Influence of Tomato Seed Flour on Dynamic Dough Rheological Properties

Figure 1 shows the mechanical spectra of dough samples formulated with TSF at different levels. Both G' and G'' moduli values increased with the increase of TSF level addition in wheat flour. The increase of the dynamic moduli can be due to the limited plasticization effect and to the TSF emulsifying properties [5] and its lipids contents which favor gluten aggregation and gives rise to a more elastic behavior [60]. The TSF proteins are not similar to the gluten proteins and thus by TSF addition the G'' values also increased indicating a more viscous behavior of the dough samples, probably due to the dilution effect on gluten proteins. A similar trend for the dynamic moduli was also found when wheat flour was substituted with grape seeds flour [42]. However, for all the analyzed samples the G'/G''

values were higher than 1, indicating a solid elastic-like behavior of dough [61] with and without TSF addition in wheat flour.

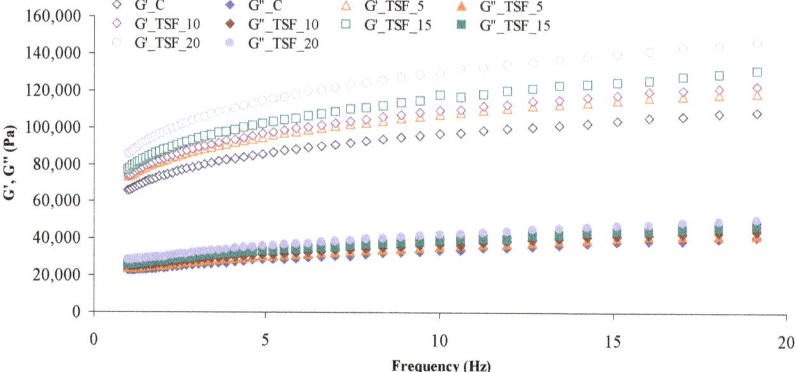

Figure 1. Mechanical spectra of control dough (C) and dough samples with different levels (5%, 10%, 15% and 20%) of tomato seeds flour (TSF). Presented data are mean values.

The parameters for power-law equations which were used to describe the dependence of moduli on the oscillation frequency are shown in Table 3. As it may be seen, the dependency of G' and G'' dynamic moduli on the oscillation frequency was well modeled by Equations (1,2) in the range of tested frequency from 1 to 20 Hz. The coefficient of determination (R^2) values was higher than 0.999 and 0.960 for G' and G'', respectively, showing that the power law model was adequate in modeling the viscoelastic properties of dough samples. The obtained values of K', K'', n' and n'' parameters adequate to G' and G'' are shown in Table 3.

Table 3. Parameters of power law models, Equations (1,2) describing the dependence of storage (G') and loss (G'') moduli on the frequency.

Sample	$G' = K' \cdot \omega^{n'}$		$G'' = K'' \cdot \omega^{n''}$	
	K' (Pa s$^{n'}$)	n'	K'' (Pa s$^{n''}$)	n''
Control	65,707.45 a (7373.07)	0.168 a (0.008)	21,489.72 a (3403.12)	0.196 a (0.008)
TSF_5	73,158.88 a (3875.74)	0.167 a (0.001)	22,721.65 a (439.82)	0.199 a (0.001)
TSF_10	73,711.68 a (2893.63)	0.172 a (0.004)	23,469.96 a (2119.38)	0.197 a (0.005)
TSF_15	77,350.93 b (2412.14)	0.177 c (0.002)	24,470.72 a (1762.34)	0.205 b (0.003)
TSF_20	85,702.71 c (7512.53)	0.182 b (0.002)	26,867.43 b (2610.38)	0.198 a (0.004)

Values in parentheses are standard deviations. [a,b,c] Values with the same letter are not significantly different according to the LSD test ($p < 0.05$).

The values of K' and K'' increased with the increase of TSF addition level in wheat flour. Significant increases ($p < 0.05$) were noticed especially between the control sample and samples in which 15% and 20% TSF were incorporated. The highest values of K' and K'' were obtained for the dough sample with the highest level of TSF addition in wheat flour. The results obtained indicated that the TSF addition led to a stronger dough compared to the control sample. This effect on the viscoelastic properties can be related to the TSF emulsifying properties which act as a filler in the dough viscoelastic matrix [62] causing strong bonds which lead to higher modulus values.

The results obtained for the G' slope represented by n' presented lower values than n'' for the G'' slope (Table 3). The TSF addition in wheat flour dough favors the binding of lipids from TSF to gluten proteins by hydrophobic interactions during mixing. Also, due to the TSF emulsifying capacity the gluten proteins charge decreased, favoring their aggregation, leading to an increase of the n' and n'' parameters values compared to the control sample. The additional level of TSF had a significant effect ($p < 0.05$) on n' at level of 15% and 20%, while on n'' only at the level of 15%. Proximate n' and n'' values ($n' = 0.22$–0.12; $n'' = 0.23$–0.181) were found by Chouaibi et al. [63] for wheat flour dough with commercial tomato products addition.

Figure 2 showed creep and recovery curves for wheat flour dough with different levels of TSF addition in wheat flour. The replacement of wheat flour with TSF caused an increase of the deformation compliance only for the sample with 15% TSF addition, as compared to control. The other blends samples showed lower compliance values during the creep test, indicating lower deformability than the control wheat flour dough. The lowest deformation compliance obtained for the sample with 20% TSF addition in wheat flour shows a lower deformability and therefore a stronger matrix for the dough structure.

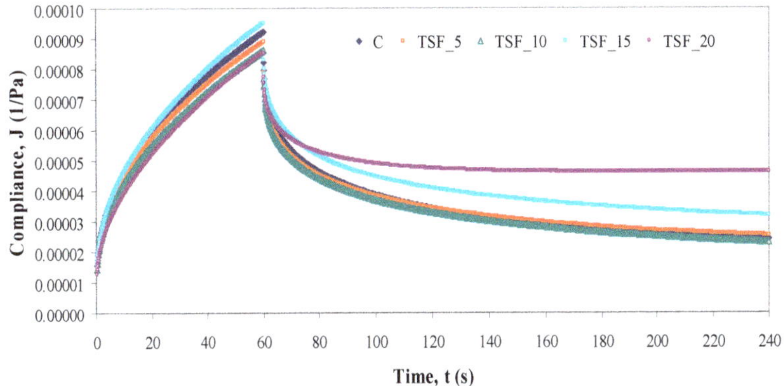

Figure 2. Creep and recovery curves of control sample (C) and dough samples with different levels (5%, 10%, 15% and 20%) of tomato seed flour (TSF). Presented data are mean values.

The experimental data of compliance were well adjusted ($R^2 > 0.97$) to the Burgers model, Equations (4,5). The parameters of Burgers model are shown in Table 4. In general, the levels of TSF addition exert a significant effect on the creep parameter values. The highest value of instantaneous compliance was observed in the case of the control sample, indicating a more elastic nature and higher recovery. An addition of TSF led to a decreased of J_{Co}, except for the sample with 15% TSF which showed a slightly increase of the instantaneous compliance. This increase can be related to a slight firmness improvement of dough structure, in according to the result reported by Miś [64]. The highest change on J_{Co} was found for the sample with 20% TSF, indicating a significantly ($p < 0.05$) decrease of the instantaneous elasticity. In respect to the retardation times obtained for the creep and recovery phase, the results showed that the TSF addition levels exhibit significance changes ($p < 0.05$) on λ parameter. The increase of λ_C values for the samples with levels higher than 5% TSF addition in wheat flour is significant, showing that the retarded elastic creep took place more slowly. The decrease of λ_R values with the increase of TSF addition level in wheat flour to 15% and 20% showed that the retarded elastic recovery took place more rapidly. A significant higher viscosity, μ_{Co} was found in the all samples with TSF addition, except the sample when 15% TSF was incorporated in wheat flour, indicating a higher dough resistance to flow deformability than the control sample. Therefore, the samples with 15% TSF addition in wheat flour presented lower opposition to deformation than the dough samples with 5%, 10% and 20% TSF. This behavior is correlated to the maximum creep compliance value (9.41) obtained for sample with 15% TSF. Compared to other samples with TSF addition, this effect can be related to the gluten dilution in dough. The decrease of gluten proteins by TSF addition at constant stress led to an increase of creep compliance, whereas the elastic contribution decreased [65].

Table 4. Parameters of Burger's model.

Sample	Creep Phase					Recovery Phase				
	$J_{Co} \cdot 10^5$ (Pa^{-1})	$J_{Cm} \cdot 10^5$ (Pa^{-1})	λ_C (s)	$\mu_{Co} \cdot 10^{-6}$ (Pa $\cdot$ s)	$J_{max} \cdot 10^5$ (Pa^{-1})	$J_{Ro} \cdot 10^5$ (Pa^{-1})	$J_{Rm} \cdot 10^5$ (Pa^{-1})	λ_R (s)	$J_r \cdot 10^5$ (Pa^{-1})	J_r/J_{max} (%)
Control	2.50 a (0.16)	8.13 a (0.04)	36.97 a (0.48)	1.48 a (0.01)	9.13 a (0.33)	2.30 a (0.20)	3.90 a (0.29)	38.19 a (3.47)	6.20 a (0.10)	67.98 a (3.50)
TSF_5	2.45 b (0.03)	7.80 b (1.07)	36.60 b (2.93)	1.60 b (0.24)	8.80 d (0.61)	2.62 b (1.28)	3.61 b (0.45)	37.10b (9.63)	6.22 b (0.83)	70.37 a (4.55)
TSF_10	2.32 e (0.12)	7.69 c (0.64)	37.95 c (0.21)	1.54 c (0.14)	8.51 a (0.54)	2.42 e (0.16)	3.60 a (0.54)	39.11 c (4.74)	6.01 a (0.39)	70.66 a (0.14)
TSF_15	2.51 c (0.36)	8.64 d (1.24)	38.56 e (0.96)	1.45 e (0.22)	9.41 b (1.32)	3.32 c (0.28)	3.49 a (0.74)	36.02d (3.66)	6.81 c (1.02)	72.27 a (0.70)
TSF_20	2.04 d (0.15)	8.41 e (1.60)	42.54 d (3.33)	1.54 d (0.26)	8.48 c (1.27)	4.51 d (2.04)	2.32 c (0.45)	22.48 e (17.21)	6.83 d (1.59)	79.56 a (6.89)

Values in parentheses are standard deviations. [abcde] Values with the same letter are not significantly different according to the LSD test ($p < 0.05$).

The recovery data obtained (Table 4) shows the elasticity variation of dough samples with TSF addition at different levels to wheat flour. The level of TSF has a significant influence ($p < 0.05$) on the instantaneous recovery compliance and the retardation time, but not on the recovery parameter (J_r/J_{max}). The control dough showed the lowest elasticity. The highest elasticity was found in the sample with 20% TSF addition, which could be associated with its higher J_{Ro} value. A higher compliance elasticity during recovery suggested higher recoverable energy stored by a more cross-linked gluten than the control sample. This fact may be due to the increased aggregation between gluten proteins in the dough system with TSF addition [66]. Among samples with TSF, the highest elasticity was found for the samples in which 15% and 20% TSF were incorporated in wheat flour.

3.5. Influence of Tomato Seed on Dough Microstructure

Dough samples with different levels of tomato seed flour addition were analyzed with EFLM (Figure 3A–E). All the samples were labeled with a solution of 1% rhodamine B and 0.5% fluorescein (FITC) in 2-methoxyethanol. In a dough system rhodamine B will label protein in red and FITC will label starch in green. Figure 3A shows a dough sample without tomato seed flour addition. In this sample, the presence of a very high amount of starch granules that were connected to the protein network which was squeezed in the starch matrix was observed. Comparing the structure images obtained for the dough samples with different levels of tomato seed addition a significant difference may be seen between the amount of starch granules and protein. When the level of tomato seed addition increased in wheat flour a less green area was seen and a more red area was present in the dough system (Figure 3B–E). This fact indicated that the starch granules became fewer and the protein content higher. Taking into account that tomato seeds present more than double the amount of proteins compared to wheat flour, it was expectable that the starch content began to decrease and the protein to increase from the dough matrix with the increase of tomato seed level addition. When low levels of tomato seed flour were incorporated in the dough system the starch granules were glued together in the matrix.

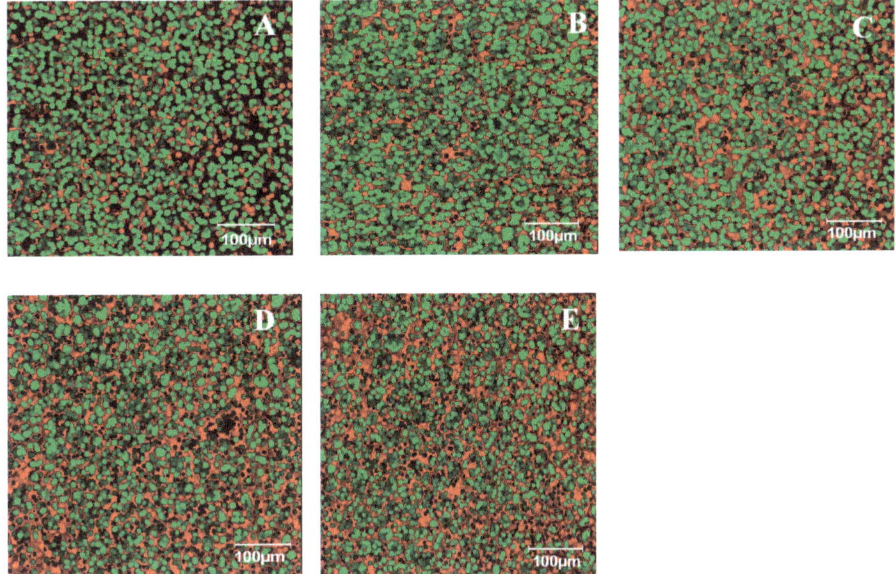

Figure 3. Microstructure taken by epifluorescence light microscopy (EFLM) of wheat dough with tomato seed flour (TSF) at different levels: 0% (**A**), 5% (**B**), 10% (**C**), 15% (**D**) and 20% (**E**). Green, starch granules; red, protein.

At high levels of tomato seed flour addition the starch granules appeared separated surrounded by the protein network. However, all the samples analyzed seemed to be homogeneous, with no black regions in the matrix, meaning that the dough did not become too weak even when higher levels of tomato seeds were incorporated. The results obtained from EFLM were in accordance with the rheological data. Even if the wheat flour substitution level with TSF was of 20% the Mixolab stability of the dough system was higher than the stability of the control samples whereas the both moduli values G' and G'' increased with the level of TSF addition increase indicating that dough with TSF was a stronger and compact one.

4. Conclusions

In conclusion, the wheat-tomato seed composite flours parameters analyzed, namely protein, fat and ash, increased with the level of tomato seed addition increase in wheat flour, indicating the fact that bread with TSF will present a higher nutritional and energetic value for consumers. According to the rheological data, all the dough samples with TSF addition were more stable and strengthened compared to the control sample presenting higher values for dough stability and development time. This fact indicates to the bakers that the dough with TSF addition presents a higher capacity to keep its shape during proofing and therefore, it may be fermented for a longer period of time. Also, dough with TSF addition will present a higher ability to retain the gas formed during the fermentation process, a fact that will influence bread porosity. The higher dough strength for the samples with TSF indicated that the bread obtained presented a more firm texture. Regarding the mixing behavior for the dough samples with TSF addition, it seemed that according to the Mixolab data the bakers should mix the composite flour for a longer period of time and must introduce less water amount in order to obtain an optimum dough consistency. At high levels of TSF addition in wheat flour, all the Mixolab parameters torques related to the starch behavior (C3, C4, C5) decreased probably due to the decrease of the starch content from the wheat–tomato seed flour samples. At low levels of addition, TSF seemed to integrate very well into the dough system increasing its starch stability paste.

Also, the FN increased when low levels of TSF were incorporated and decreased when high levels of TSF were added in dough samples. This, along with starch behavior recorded with the Mixolab device, showed that the addition of α amylase may be recommended for wheat flour with high FN values without affecting in a negatively way the dough stability, a fact that will lead to good bakery products quality. It must been mention that when high levels of TSF were added the C5 and C5-4 begin to decrease, favoring less starch retrogradation, indicating the fact that bakery products obtained with TSF addition will present a higher shelf life than the control sample. The dynamic rheological data showed that with the increase of TSF addition level in wheat flour the dough samples presented higher viscoelastic solid properties. Creep–recovery tests showed that TSF addition in wheat flour led to dough with higher resistance to deformation, smaller creep strain values and higher recovery, indicating stronger dough with greater elasticity. This indicated that during the technological process the dough samples with TSF addition presented the capability to maintain their form, a fact that will create the possibility to prolong the fermentation phase due to a higher ability to retain the gas formed during this process. Also, after baking the bakery products may present higher loaf volumes and porosity due to the fact that by prolonging the fermentation process the amount of the gas released will increase and may be retained by the dough system due to it high elasticity. The results obtained from EFLM were in accordance with the rheological data. All the analyzed samples were homogeneous, with a compact dough matrix, showing that the dough samples were strong even when high levels of TSF were incorporated. According to the data obtained, we concluded that tomato seed, as a whole flour form, can be used in bread making by blending with wheat flour even at high levels of up to 20%, as an alternative for developing new bakery products. The TSF addition increased the dough strength which makes us recommend its use for wheat flour of a good quality for bread making with a high content of wet gluten. In the case of its addition in wheat flour with high FN values we recommend the addition of α amylase in the dough system, a fact that will increase the amount of the gas formed in the dough system, gas that dough with TSF addition is capable of retaining.

Author Contributions: Contributed equally to the study design, collection of data, development of the sampling, analyses, interpretation of results, and preparation of the paper, S.M. and G.G.C. All authors read and approved the final manuscript.

Acknowledgments: This work was supported from contract no. 18PFE/16.10.2018 funded by the Ministry of Research and Innovation within Program 1—Development of national research and development system, Subprogram 1.2—Institutional Performance—RDI excellence funding projects.

Conflicts of Interest: The authors declare no conflict of interest.

References

1. Adalid, A.M.; Roselló, S.; Nuez, F. Evaluation and selection of tomato accessions (*Solanum* section *Lycopersicon*) for content of lycopene, β-carotene and ascorbic acid. *J. Food Compos. Anal.* **2010**, *23*, 613–618. [CrossRef]
2. Bajerska, J.; Chmurzynska, A.; Mildner-Szkudlarz, S.; Drzymala-Czyz, S. Effect of rye bread enriched with tomato pomace on fat absorption and lipid metabolism in rats fed a high-fat diet. *J. Sci. Food Agric.* **2015**, *95*, 1918–1924. [CrossRef] [PubMed]
3. Vinkovic Vrcek, I.; Samobor, V.; Bojic, M.; Medic-Saric, M.; Vukobratovic, M.; Erhatic, R.; Horvat, D.; Matotan, Z. The effect of grafting on the antioxidant properties of tomato (*Solanum lycopersicum* L.). *Span. J. Agric. Res.* **2011**, *9*, 844–851. [CrossRef]
4. Maršić, N.K.; Gasperlin, L.; Abram, V.; Budic, M.; Vidrih, R. Quality parameters and total phenolic content in tomato fruits regarding cultivar and microclimatic conditions. *Turk. J. Agric. For.* **2011**, *35*, 185–194.
5. Shao, D.; Atungulu, G.; Pan, Z.; Yue, T.; Zhang, A.; Fan, Z. Characteristics of isolation and functionality of protein from tomato pomace produced with different industrial processing methods. *Food Bioprocess Technol.* **2014**, *7*, 532–541. [CrossRef]
6. Borguini, R.G.; Ferraz da Silva Torres, E.A. Tomatoes and tomato products as dietary sources of antioxidants. *Food Rev. Int.* **2009**, *25*, 313–325. [CrossRef]
7. Ray, R.C.; El Sheikha, A.F.; Panda, S.H.; Montet, D. Anti-oxidant properties and other functional attributes of tomato: An overview. *Int. J. Food Ferment. Technol.* **2011**, *1*, 139–148.

8.   Zuorro, A.; Fidaleo, M.; Lavecchia, R. Enzyme-assisted extraction of lycopene from tomato processing waste. *Enzym. Microb. Technol.* **2011**, *49*, 567–573. [CrossRef]

9.   Del Valle, M.; Camara, M.; Torija, M.E. The nutritionaland functional potential of tomato by-products. *Acta Hortic.* **2007**, *758*, 165–172. [CrossRef]

10.   Celma, A.R.; Cuadros, F.; López-Rodríguez, F. Characterisation of industrial tomato by-products from infrared drying process. *Food Bioprod. Process.* **2009**, *87*, 282–291. [CrossRef]

11.   Migalatiev, O. Optimisation of operating parameters for supercritical carbon dioxide extraction of lycopene from industrial tomato waste. *Ukr. Food J.* **2017**, *6*, 698–716. [CrossRef]

12.   Sarkar, A.; Kaul, P. Evaluation of Tomato Processing By-Products: A Comparative Study in a Pilot Scale Setup. *J. Food Process Eng.* **2014**, *37*, 299–307. [CrossRef]

13.   Kumar, K.; Yadav, A.N.; Kumar, V.; Vyas, P.; Dhaliwal, H.S. Food waste: A potential bioresource for extraction of nutraceuticals and bioactive compounds. *Bioresour. Bioprocess.* **2017**, *4*, 18. [CrossRef]

14.   Knoblich, M.; Anderson, B.; Latshaw, D. Analyses of tomato peel and seed byproducts and their use as a source of carotenoids. *J. Sci. Food Agric.* **2005**, *85*, 1166–1170. [CrossRef]

15.   Nour, V.; Ionica, M.E.; Trandafir, I. Bread enriched in lycopene and other bioactive compounds by addition of dry tomato waste. *J. Food Sci. Technol.* **2015**, *52*, 8260–8267. [CrossRef] [PubMed]

16.   Kalogeropoulos, N.; Chiou, A.; Pyriochou, V.; Peristeraki, A.; Karathanos, V.T. Bioactive phytochemicals in industrial tomatoes and their processing byproducts. *LWT-Food Sci. Technol.* **2012**, *49*, 213–216. [CrossRef]

17.   Mehta, D.; Prasad, P.; Sangwan, R.S.; Yadav, S.K. Tomato processing byproduct valorization in bread and muffin: Improvement in physicochemical properties and shelf life stability. *J. Food Sci. Technol.* **2018**, *55*, 2560–2568. [CrossRef]

18.   Papaioannou, E.H.; Karabelas, A.J. Lycopene recovery from tomato peel under mild conditions assisted by enzymatic pre-treatment and non-ionic surfactants. *Acta Biochim. Pol.* **2012**, *59*, 71–74. [CrossRef]

19.   Stajčić, S.; Ćetković, G.; Čanadanović-Brunet, J.; Djilas, S.; Mandić, A.; Četojević-Simin, D. Tomato waste: Carotenoids content, antioxidant and cell growth activities. *Food Chem.* **2015**, *172*, 225–232. [CrossRef]

20.   Gupta, G.; Parihar, S.S.; Ahirwar, N.K.; Snehi, S.K.; Singh, V. Plant growth promoting rhizobacteria (PGPR): Current and future prospects for development of sustainable agriculture. *J. Microb. Biochem. Technol.* **2015**, *7*, 096–102.

21.   Sogi, D.S.; Garg, S.K.; Bawa, A.S. Functional properties of seed meals and protein concentrates from tomato-processing waste. *J. Food Sci.* **2002**, *67*, 2997–3001. [CrossRef]

22.   Sogi, D.S.; Sidhu, J.S.; Arora, M.S.; Garg, S.K.; Bawa, A.S. Effect of tomato seed meal supplementation on the dough and bread characteristics of wheat (PBW 343) flour. *Int. J. Food Prop.* **2002**, *5*, 563–571. [CrossRef]

23.   Kramer, A.; Kwee, W.H. Functional and nutritional properties of tomato protein concentrates. *J. Food Sci.* **1977**, *42*, 207–211. [CrossRef]

24.   Sogi, D.S.; Bhatia, R.; Garg, S.K.; Bawa, A.S. Biological evaluation of tomato waste seed meals and protein concentrate. *Food Chem.* **2005**, *89*, 53–56. [CrossRef]

25.   Mechmeche, M.; Kachouri, F.; Chouabi, M.; Ksontini, H.; Setti, K.; Hamdi, M. Optimization of extraction parameters of protein isolate from tomato seed using response surface methodology. *Food Anal. Methods* **2017**, *10*, 809–819. [CrossRef]

26.   Latlief, S.J.; Knorr, D. Tomato seed protein concentrates: Effects of methods of recovery upon yield and compositional characteristics. *J. Food Sci.* **2010**, *48*, 1583–1586. [CrossRef]

27.   Majzoobi, M.; Ghavi, F.S.; Farahnaky, A.; Jamalian, J.; Mesbahi, G. Effect of tomato pomace powder on the physicochemical properties of flat bread (barbari bread). *J. Food Process. Preserv.* **2011**, *35*, 247–256. [CrossRef]

28.   Brodowski, D.; Geisman, J.R. Protein content and amino acid composition of protein of seeds from tomatoes at various stages of ripeness. *J. Food Sci.* **1980**, *45*, 228–229. [CrossRef]

29.   Rahma, E.H.; Moharram, Y.G.; Mostafa, M.M. Chemical Characterisation of Tomato Seed Protein (Var. Pritchard). *Egypt. J. Food Sci.* **1986**, *14*, 221–230.

30.   Bhat, M.A.; Ahsan, H. Physico-chemical characteristics of cookies prepared with tomato pomace powder. *J. Food Process. Technol.* **2016**, *7*. [CrossRef]

31.   Savadkoohi, S.; Hoogenkamp, H.; Shamsi, K.; Farahnaky, A. Color, sensory and textural attributes of beef frankfurter, beef ham and meat-free sausage containing tomato pomace. *Meat Sci.* **2014**, *97*, 410–418. [CrossRef] [PubMed]

32. Wang, Q.; Xiong, Z.; Li, G.; Zhao, X.; Wu, H.; Ren, Y. Tomato peel powder as fat replacement in low-fat sausages: Formulations with mechanically crushed powder exhibit higher stability than those with airflow ultra-micro crushed powder. *Eur. J. Lipid Sci. Technol.* **2016**, *118*, 175–184. [CrossRef]

33. Grassino, A.N.; Halambek, J.; Djakovic, S.; Brncic, S.R.; Dent, M.; Grabaric, Z. Utilization of tomato peel waste from canning factory as a potential source for pectin production and application as tin corrosion inhibitor. *Food Hydrocoll.* **2016**, *52*, 265–274. [CrossRef]

34. Carlson, B.L.; Knorr, D.; Watkins, T.R. Influence of tomato seed addition on the quality of wheat-flour breads. *J. Food Sci.* **1981**, *46*, 1029–1031. [CrossRef]

35. Mironeasa, S.; Codina, G.G.; Oroian, M.A. Bread quality characteristics as influenced by the addition of tomato seed flour. *Bull. Univ. Agric. Sci. Vet. Med.* **2016**, *73*, 77–84. [CrossRef]

36. Mironeasa, S.; Codină, G.G.; Mironeasa, C. Effect of Composite Flour Made from Tomato Seed and Wheat of 650 Type of a Strong Quality for Bread Making on Bread Quality and Alveograph Rheological Properties. *Int. J. Food Eng.* **2018**, *4*, 22–26. [CrossRef]

37. Kowalczewski, P.Ł.; Walkowiak, K.; Masewicz, Ł.; Bartczak, O.; Lewandowicz, J.; Kubiak, P.; Baranowska, H.M. Gluten-Free Bread with Cricket Powder—Mechanical Properties and Molecular Water Dynamics in Dough and Ready Product. *Foods* **2019**, *8*, 240. [CrossRef]

38. Korus, J.; Juszczak, L.; Ziobro, R.; Witczak, M.; Grzelak, K.; Sójka, M. Defatted strawberry and blackcurrant seeds as functional ingredients of gluten-free bread. *J. Texture Stud.* **2012**, *43*, 29–39. [CrossRef]

39. Steffe, J.F. *Rheological Methods in Food Process Engineering*; Freeman Press: East Lansing, MI, USA, 1996; pp. 294–348.

40. Mironeasa, S.; Iuga, M.; Zaharia, D.; Mironeasa, C. Rheological analysis of wheat flour dough as influenced by grape peels of different particle sizes and addition levels. *Food Bioprocess Technol.* **2019**, *12*, 228–245. [CrossRef]

41. Moreira, R.; Chenlo, F.; Torres, M.D. Rheology of gluten-free doughs from blends of chestnut and rice flours. *Food Bioprocess Technol.* **2013**, *6*, 1476–1485. [CrossRef]

42. Iuga, M.; Mironeasa, C.; Mironeasa, S. Oscillatory rheology and creep-recovery behaviour of grape seed-wheat flour dough: Effect of grape seed particle size, variety and addition level. *Bull. Univ. Agric. Sci. Vet. Med. Cluj-Napoca Food Sci. Technol.* **2019**, *76*, 37–45. [CrossRef]

43. Abebe, W.; Ronda, F.; Villanueva, M.; Collar, C. Effect of tef [Eragrostis tef (Zucc.) Trotter] grain flour addition on viscoelastic properties and stickiness of wheat dough matrices and bread loaf volume. *Eur. Food Res. Technol.* **2015**, *241*, 469–478. [CrossRef]

44. Peighambardoust, S.H.; Dadpour, M.R.; Dokouhaki, M. Application of epifluorescence light microscopy (EFLM) to study the microstructure of wheat dough: A comparison with confocal scanning laser microscopy (CSLM) technique. *J. Cereal Sci.* **2010**, *51*, 21–27. [CrossRef]

45. Codina, G.G.; Mironeasa, S. Influence of mixing speed on dough microstructure and rheology. *Food Technol. Biotechnol.* **2013**, *51*, 509–519.

46. Codină, G.G.; Mironeasa, S. Use of response surface methodology to investigate the effects of brown and golden flaxseed on wheat flour dough microstructure and rheological properties. *J. Food Sci. Technol.* **2016**, *53*, 4149–4158. [CrossRef]

47. Ionescu, V.; Stoenescu, G.; Vasilean, I.; Aprodu, I.; Banu, I. Comparative evaluation of wet gluten quantity and quality through different methods. *Food Technol.* **2010**, *34*, 44–48.

48. Struyf, N.; Verspreet, J.; Courtin, C.M. The effect of amylolytic activity and substrate availability on sugar release in non-yeasted dough. *J. Cereal Sci.* **2016**, *69*, 111–118. [CrossRef]

49. Kundu, H.; Grewal, R.B.; Goyal, A.; Upadhyay, N.; Prakash, S. Effect of incorporation of pumpkin (Cucurbita moshchata) powder and guar gum on the rheological properties of wheatflour. *J. Food Sci. Technol.* **2014**, *51*, 2600–2607. [CrossRef]

50. Chin, N.L.; Goh, S.K.; Rahman, R.A.; Hashim, D.M. Functional effect of fullyhydrogenated palm oil-based emulsifiers on baking performance of whitebread. *Int. J. Food Eng.* **2007**, *3*. [CrossRef]

51. Majzoobi, M.; Mesbahi, G.; Farahnaky, A.; Jamalian, J.; Sariri, F. Effects of tomato pulp and sugar beet pulp powders on the farinograph properties of bread dough. *Food Sci. Technol.* **2011**, *8*, 1–9.

52. Guo, H.; Lin, Y.H.; Wan, S.W.; Guo, S.Y. Application of emulsifier in baking food. Mod. *Food Sci. Technol.* **2006**, *22*, 297–298.

53. Pareyt, B.; Finnie, S.M.; Putseys, J.A.; Delcour, J.A. Lipids in bread making. Sources, intercations and impact on bread quality. *J. Cereal Sci.* **2011**, *54*, 266–279. [CrossRef]

54. Piyakina, G.A.; Maksimov, V.V.; Yunusov, T.S. Some properties of the protein fractions of tomato seeds. *Chem. Nat. Compd.* **1998**, *34*, 492–495. [CrossRef]

55. Sarkar, A.; Kamaruddin, H.; Bentley, A.; Wang, S. Emulsion stabilization by tomato seed protein isolate: Influence of pH, ionic strength and thermal treatment. *Food Hydrocoll.* **2016**, *57*, 160–168. [CrossRef]

56. Savadkoohi, S.; Farahnaky, A. Dynamic rheological and thermal study of the heat-induced gelation of tomato-seed proteins. *J. Food Eng.* **2012**, *113*, 479–485. [CrossRef]

57. Moza, J.; Gujral, S. Influence of barley non-starchy polysaccharides on selected quality attributes of sponge cakes. *LWT-Food Sci. Technol.* **2017**, *85*, 252–261. [CrossRef]

58. Gujral, H.S.; Sharma, B.; Khatri, M. Influence of replacing wheat bran with barley bran on dough rheology, digestibility and retrogradation behavior of chapatti. *Food Chem.* **2018**, *240*, 1154–1160. [CrossRef]

59. Wang, S.; Li, C.; Copeland, L.; Niu, Q.; Wang, S. Starch retrogradation: A comprehensive review. *Compr. Rev. Food Sci. Food Saf.* **2015**, *14*, 568–585. [CrossRef]

60. Watanabe, A.; Yokomizo, K.; Eliasson, A.-C. Effect of physical states of nonpolar lipids on rheology, ultracentrifugation, and microstructure of wheat flour dough. *Cereal Chem.* **2003**, *80*, 281–284. [CrossRef]

61. Guadarrama-Lezama, A.Y.; Carillo-Navas, H.; Vernon-Carter, E.J.; Alverez-Ramirez, J. Rheological and thermal properties of dough and textural and microstructural features of bread obtained from nixtamalized corn/wheat flour blends. *J. Cereal Sci.* **2016**, *69*, 158–165. [CrossRef]

62. Uthayakumaran, S.; Newberry, M.; Phan-Thien, N.; Tanner, R. Small and large strain rheology of wheat gluten. *Rheol. Acta* **2002**, *41*, 162–172. [CrossRef]

63. Chouaibi, M.; Rezig, I.; Boussaid, A.; Hamdi, S. Insoluble tomato-fiber effect on wheat dough rheology and cookies'quality. *Ital. J. Food Sci.* **2018**, *31*. [CrossRef]

64. Miś, A. Interpretation of mechanical spectra of carob fibre and oat wholemeal-enriched wheat dough using non-linear regression models. *J. Food Eng.* **2011**, *102*, 369–379. [CrossRef]

65. Wang, F.C.; Sun, X.S. Creep-recovery of wheat flour doughs and relationship to other physical dough tests and breadmaking performance. *Cereal Chem.* **2002**, *70*, 567–571. [CrossRef]

66. Chompoorat, P.; Ambardekar, A.; Mulvaney, S.; Rayas-Duarte, P. Rheological characteristics of gluten after modified by DATEM, ascorbic acid, urea and DTT using creep-recovery test. *J. Mod. Phys.* **2013**, *4*, 1–8. [CrossRef]

MDPI

St. Alban-Anlage 66

4052 Basel

Switzerland

Tel. +41 61 683 77 34

Fax +41 61 302 89 18

www.mdpi.com

Foods Editorial Office

E-mail: foods@mdpi.com

www.mdpi.com/journal/foods